SpringerBriefs in Business

More information about this series at http://www.springer.com/series/8860

Emanuela Delbufalo

Agency Theory and Sustainability in the Global Supply Chain

Springer

Emanuela Delbufalo
Department of Human Science
European University of Rome
Rome
Italy

ISSN 2191-5482 ISSN 2191-5490 (electronic)
SpringerBriefs in Business
ISBN 978-3-319-72792-9 ISBN 978-3-319-72793-6 (eBook)
https://doi.org/10.1007/978-3-319-72793-6

Library of Congress Control Number: 2018937703

Printed on acid-free paper

This Springer imprint is published by the registered company Springer International Publishing AG part of Springer Nature
The registered company address is: Gewerbestrasse 11, 6330 Cham, Switzerland

Preface

As corporations have increased outsourcing and global sourcing from low-cost countries in Eastern Europe and the Far East, a number of different stakeholders have expressed growing concerns about the ethical, social and environmental impacts of production and consumption in global supply chains (Andersen and Skjøtt-Larsen 2009). At the same time, supply chain management literature has shown a renewed interest in issues related to environmental management, green supply chains, sustainable sourcing and social responsibility (Lee and Kim 2009). As firms increasingly rely on extended networks of suppliers to produce and deliver products and services to customers, the concepts of corporate social responsibility (CSR) and sustainable development are subject to a transformation. They are no longer the individual firm's domain; increasingly, they encompass all the activities developed by multiple actors along with the entire supply chain (e.g., Amaeshi et al. 2008).

Some recent events and business scandals[1] have demonstrated that, in the long term, for buyers from developed countries, it is not beneficial to take advantage of the poor labour conditions of supply firms in developing countries and their relative lack of social and environmental standards. The detrimental performance of suppliers is likely to impact, in a severe way, on the sustainability of business strategies and the reputation of the buyers (often multinational corporations, MNCs) (Jiang 2009). In order to respond to the criticism linked to poor social, ethical and environmental practices in global supply chains, a growing number of MNCs have engaged in collaborative initiatives developed to design and enforce sustainability practices towards a broad number of stakeholders.

One of the main purposes of MNCs' collaborative efforts is the development of a shared code of conduct and a set of monitoring initiatives to manage the relationships with suppliers and assure their compliance with ethically and socially responsible practices (Delbufalo and Bastl 2018; Mele and Schepers 2013).

[1] See, for example, Clean Clothes Campaign (2016)

Moreover, the collaborative efforts between MNCs have also been seen as a way to establish common policies and action plans, design and encourage the development of joint best practices, and in general avoid exposing suppliers to potentially conflicting requirements (Aguinis and Glavas 2012). In doing so, such collaboration aims to reduce the overall costs and the high degree of uncertainty typically present in the implementation of CSR practices on a large scale (Arya and Salk 2006).

The aim of this study is to analyse the impacts of corporate responsibility collaboration among MNCs on the suppliers' degree of compliance with sustainability requirements. Anecdotal evidence shows that—despite the growing efforts of MNCs in extending the concept and practice of sustainability to other actors in the supply chain—poor suppliers' social and environmental performance and audit frauds in developing countries have escalated (Clean Clothes Campaign 2016; Egels-Zandén 2014). Suppliers' opportunistic behaviour raises questions about what buyers from developed countries should do to enforce CSR implementation (Jiang 2009). This study intends to answer these questions by adopting an agency theory perspective, which is specifically focused on the relationship between two agents that are engaged in cooperative behaviour but have differing goals and asymmetric information (Eisenhardt 1989).

This first chapter is dedicated to a literature review of agency theory in supply chain management contexts. The review presents the state of the art of research on the topic, discusses the areas in which agency theory has been more usefully adopted and shows its more valuable explanatory power. In particular, agency theory provides a solid explanation for the motivations for opportunistic behaviours in buyer–supplier relationships and makes suggestions for reducing their frequency and negative returns. The second chapter addresses the antecedents of suppliers' (mis)conduct along with the role and function of buyers' codes of conduct in global supply chains. The chapter also presents the main characteristics of multi-stakeholder collaborative initiatives developed to respond to multiple stimuli coming from customers and society. The third chapter discusses the contribution of corporate responsibility collaborations among competitors. By adopting an agency theory perspective, the study explains the origin of agency problems in global supply chains and the negative impact of opportunistic behaviours on sustainable development. After that, the study addresses the effects of collaboration among competitors on the degree of suppliers' compliance and analyses how this collaboration shapes the code of conduct's monitoring system and relationship management. The study enriches the theoretical arguments with some notable experiences of industry-specific collaborations for sustainable supply chain management and shared code of conduct implementation.

Rome, Italy Emanuela Delbufalo

References

Aguinis, H., & Glavas, A. (2012). What we know and don't know about Corporate Social Responsibility: A review and research agenda. *Journal of Management*, *38*(4), 932–968.

Amaeshi, K.M., Osuji, O.K., Nnodim, P. (2008). Corporate Social Responsibility in supply chains of global brands: A boundaryless responsibility? Clarifications, exceptions and implications. *Journal of Business Ethics*, *81*(1), 223–234.

Andersen, M., & Skjøtt-Larsen, T. (2009). Corporate social responsibility in global supply chains. *Supply Chain Management: An International Journal*, *14*(2), 75–86.

Arya, B., & Salk, J. (2006). Cross-sector alliance learning and effectiveness of voluntary codes of corporate social responsibility. *Business Ethics Quarterly*, *16*(2), 211–234.

Clean Clothes Campaign (2016). Dangerous delays on workers safety. https://cleanclothes.org/resources/publications/dangerous-delays-on-workersafety/view.

Delbufalo, E., & Bastl, M. (2018). Multi-principal collaboration and supplier's compliance with codes-of-conduct. *International Journal of Logistics Management*, forthcoming.

Egels-Zandén, N. (2014). Revisiting supplier compliance with MNC codes of conduct: Recoupling policy and practice at Chinese toy suppliers. *Journal of Business Ethics*, *119*, 59–75.

Eisenhrdt, K.M. (1989). Agency theory: An assessment and review. *Academy of Management Review*, *14*(1), 57–74.

Jiang, B. (2009). Implementing supplier codes of conduct in global supply chains: Process explanations from theoretic and empirical perspectives. *Journal of Business Ethics*, *85*, 77–92.

Lee, K-H., & Kim, J.W. (2009). Current status of CSR in the realm of supply management: the case of the Korean electronics industry. *Supply Chain Management: An International Journal*, *14*(2), 138–148.

Mele, V. & Schepers, D.H. (2013). E pluribus unum? Legitimacy issues and multi-stakeholder codes of conduct. *Journal of Business Ethics*, *118*, 561–576.

Contents

Chapter 1
Agency Theory and Supply Chain Management: A Literature Review

Abstract The chapter is dedicated to the literature review on agency theory in supply chain management contexts. After an overview on the original formulation of agency theory and a brief description of the most recent development, the review presents the state of the art of theoretical and empirical research on the topic and discusses the areas in which agency theory has been more usefully adopted. The study also shows the explanatory power of the theory for understanding the role of opportunism in supply chain relationships.

Keywords Agency theory · Opportunism · Review · Supply chain

1.1 An Overview of the Agency Theory

Agency theory, in its modern form, largely originates from the works of Mitnick (1973) and Ross (1973) and embraces the areas of political science and economics. Nevertheless, many previous works on economic risk analysis and individual–group goal incongruence have contributed to its evolution (e.g. Arrow 1985; Wilson 2006). In the management literature, the tenets of agency theory have been adapted and developed by Eisenhardt (1989) and Jensen and Meckling (1976).

In agency relationships, one party (the principal) delegates a task/activity to another party (the agent). Agency theory explicitly addresses under which contractual arrangements the relationship between principal and agent operates most efficiently. The notion of the contract is used as a metaphor to describe agency relationships (Jensen and Meckling 1976). The contract is designed considering the outcome or behaviour of the agent (Eisenhardt 1989). In agency relationships, typically, the principal seeks to minimise the agency costs—such as specifying, rewarding and monitoring the agent's behaviour—while the agent works towards maximising rewards and reducing the principal's control (Fayezi et al. 2012). The agency theory considers some basic assumptions about people (e.g. self-interest, bounded rationality, risk aversion), organisations (e.g. goal conflict among members) and information (e.g. information is a commodity which can be purchased) (Eisenhardt 1989). In each principal–agent relationship, the basic question becomes: "Is a behaviour-oriented con-

E. Delbufalo, *Agency Theory and Sustainability in the Global Supply Chain*, SpringerBriefs in Business, https://doi.org/10.1007/978-3-319-72793-6_1

tract (e.g., salaries, hierarchical governance) more efficient than an outcome-oriented contract (e.g., commissions, stock options, transfer of property rights, market governance)?" (Eisenhardt 1989: p. 58). Efficient management of agency problems, such as information acquisition, preference mismatch (or conflict of interest), effort (or moral hazard) and capability (or adverse selection), mainly associated with the agent (Fleisher 1991), is also imperative to any principal–agent relationship (Fayezi et al. 2012). It is important to note that the original formulation of agency theory defines efficiency from the principal's point of view. This assumption is based on the recognition of the principal as the dominant party in the relationship. Thus, an efficient contract is "one that brings about the best possible outcome for the principal given the constraints imposed by the situation, rather than one that maximises the joint utility of both principal and agent" (Bergen et al. 1992).

The recent development of agency theory can be categorised into two streams of inquiry: the principal–agent research and the positivist agency theory. The principal–agent approach assumes that the principal and agent will attempt to maximise their positions through individual interpretation of the contract. This approach is grounded in economic research and widely adopts mathematical modelling for relationship building (Eisenhardt 1989). In contrast, the positivist agency theory adopts a descriptive logic to interpret the real-world behaviours by using the agency theory logic and assumptions. Even if both approaches have been widely discussed in academic settings, "the mathematical and non-empirical orientation of principal-agent research have been the primary causes of this stream of research stagnating academically, particularly in organisational research" (Fayezi et al. 2012: p. 557).

Positivist agency theory has been largely extended in order to synthesise different disciplines—such as political science, expert agency and sociology—into a comprehensive framework. It aims to explain the main features of business and government relationships and offer suggestions to improve their efficiency and effectiveness (Shapiro 2005). This stream of research also provides explanations for the minimisation of a number of economic and business issues, such as separation of control from ownership (Fama and Jensen 1983; Halldórsson and Skjøtt-Larsen 2006) and the non-rational behaviour of agents (Jensen and Meckling 1976).

A principal faces two kinds of (agency) problems when entering a relationship with an agent: pre-contractual and post-contractual. The first arises before the principal decides to offer an agent a contract. The major issues here are: (a) the screening of the agent's abilities and intentions in order to find the most appropriate characteristics for the required task and (b) the definition of the strategy adopted by the principal to detect and control such abilities. The post-contractual problems emerge after the principal and agent engage in a contractual relationship. The major issues here are: (1) the definition of the incentive structure able to motivate the agent to behave in a manner consistent with the principal's goal and (2) the definition of the principal's information strategy essential for the evaluation and reward of the agent's performance. Compensations or incentive programmes for employees are typical outcomes of such decisions (Bergen et al. 1992). The literature on agency theory refers to contractual issues as problems of "adverse selection or hidden information" and post-contractual issues as problems of "moral hazard or hidden action" (Arrow 1985; Hart and Holmström 1987).

1.1.1 Hidden Information and Action Models

In order to address the issues related to misrepresentation of ability (adverse selection) and lack of effort (moral hazard) of the agent, agency theory research proposes two different approaches, respectively, designed to help the principal in designing the appropriate contract: the hidden information and the hidden action models (Bergen et al. 1992; Fayezi et al. 2012). Both models work under the informative asymmetry assumption according to which the principal is assumed to know the nature and characteristics of the task as well as the abilities needed for the agent to successfully perform the task.

Hidden information models recognise the impossibility for the principal to know if the agent has the required abilities to accomplish the contractual obligations and if he/she has the intention to do so. The principal can attempt to overcome these problems in at least three ways: screening, examining signals from the potential agents or providing opportunities for self-selection (Bergen et al. 1992). The screening strategy refers to the principal's attempt to unveil the actual abilities and attributes of the potential agent by using different means such as, for example, personal interviews and references. Alternatively, the principal might hire the agent on the basis of the imperfect information provided by the agent's signals and then learn from subsequent experience the true abilities and intentions (Bergen et al. 1992). It follows that screening is most likely to be an efficient solution to the hidden information problem when it is relatively easy to obtain information about the agent.[1] In the signalling strategy, the agent (who knows his/her own capabilities) engages in actions with the only aim being to communicate (i.e. signal) to the principal that he/she has the abilities required to perform the principal's task successfully. For such actions to yield a worthwhile signal from the agent's point of view, however, "it must meet the criteria of individual rationality and incentive compatibility. The action must make the agent better off than he/she would be by doing nothing and the benefits to be derived from the action must exceed its cost" (Bergen et al. 1992: p. 6). However, the principal is not a merely passive receiver of signals from different potential agents. He/she might also proactively construct choices or provide opportunities that enable potential agents to signal their abilities and willingness to expend effort through self-selection. A training programme for new recruits is a good example of opportunities for agents' self-selection.

---

[1] "When screening costs and the costs of learning by experience are both high, the principal may have no choice but to assume that all available agents are undesirable types and to adjust the contract offer accordingly. This breakdown, known as the market for lemons, suggests that hidden information problems can sometimes result in losses to both the principal and the agent" (Bergen et al. 1992: p. 6).

Hidden action models help the principal in choosing an appropriate contract to mitigate moral hazard and motivate the agent to take the desired action given three basic assumptions: (1) principal and agent are assumed to be motivated by self-interest; (2) the principal takes decisions under conditions of incomplete information; and (3) the relationship's outcomes are partly determined by environmental factors such as economic conditions, technological changes and environmental uncertainty in addition to the agent's behaviour. According to Bergen et al. (1992), another assumption undermines the efficiency of principal–agent relationships: the different risk preferences of the two actors. In fact, "in most hidden action models the principal is assumed to be risk neutral—or at least more neutral than the agent—whereas agents are typically assumed to be risk averse"[2] (Bergen et al. 1992: p. 4). Different risk preferences between principal and agent lead to conflicting goals, thus orienting the two actors to different strategies and courses of action (Arrow 1985). The risk preference assumption suggests that the principal has more possibilities to diversify his/her investments than the agent. The latter is supposed to be more dependent on the principal and thus more inclined to inappropriate behaviours. However, anecdotal evidence suggests that agents are often prepared to accept unbalanced risks and have a number of different choices to reduce the dependence on—and thus the control from—the principal (see, for example, Basov and Bardsley 2005). The arguments above are particularly important in supply chain contexts, where factors relating to commitment and trust compensate for goal incongruence and different risk perceptions of the actors involved, and also outweigh in many situations contractual relationships and formal governance (Fayezi et al. 2012; O'Loughlin and Clements 2007).

1.2 Descriptive Analysis of the Literature Review on Agency Theory in Supply Chain Management Contexts

The need for an agency theory perspective in supply chain management contexts arises from the consideration of several shortcomings of Transaction Cost Economics (TCE) in explaining supply chain dynamics (see Fayezi et al. 2012 for additional details). Essentially, agency theory is supposed to provide a "mechanism that may be used to explain how players (both independently and as a collective) within the [supply chain] respond to transaction cost dilemmas where rational and non-rational behaviour occurs" (p. 558). According to Stock (1997), agency theory may provide alternative explanations for a number of issues relevant for both research and practice. Some of these are: (1) the development of inter- and intra-organisational relationships; (2) the maintenance of complex relationships between suppliers and customers; (3) the effects of supply chain decisions concerning risk sharing, conflictual relationships and power; and (4) identifying the costs and benefits of make-or-buy

[2]Some models have relaxed the assumption of the risk averse agent (e.g., Harris and Raviv 1979), thus extending the external validity of hidden action models.

decisions (Logan 2000). In addition, there have been arguments in support of the combination of TCE and agency theory outside the supply chain management context (Williamson 1988) but additional research is necessary to show the real potential of such a combination.

In spite of its recognised explanatory power, in the supply chain management literature only a limited number of studies have adopted agency theory as a theoretical view.

Through a documentary research using electronic databases and a data reduction process, we have updated the literature review by Fayezi et al. (2012) on agency theory and its application in supply chain management contexts. A final list of 22 articles has been identified as relevant for the literature review's objective, and these are analysed through a detailed content analysis. The literature survey has been undertaken using the following online databases: Emerald, Science Direct, Inderscience and ABI/Inform Global ProQuest. As suggested by Fayezi et al. (2012) and Delbufalo (2012), a three-stage process using data reduction has been adopted. The selected articles have been included/excluded in the process using keywords, title, abstract and conclusion. The inclusion criteria used in the literature screening process aim to unveil whether supply chain management (and related issues such as manufacturing, procurement and logistics) and agency theory arguments are contextually discussed in the abstract and conclusion of the selected articles. A synthesis of the content analysis is reported in Table 1.1.

Figure 1.1 shows the journals with agency theory applications within supply chain management contexts. Only a few outlets collected more than one study addressing the review topic. The *International Journal of Logistics Management* was the only journal hosting three of the selected articles. The information collected in the articles was then used to identify the main themes for agency theory and supply chain management literature shown below.

1.3 Content Analysis and Stream of Research

From the analysis described above, a number of important streams of research and issues can be identified. In general, the literature review suggests two main considerations: (1) agency theory has been adopted in supply chain management research to address various areas such as procurement, manufacturing, logistics and new product development (NPD) and (2) positivist agency theory plays a dominant role in supply chain management, being adopted in the vast majority of the selected studies. The latter consideration corroborates the thesis according to which positivist agency theory can be more easily adopted by organisational researchers "mainly due to its non-mathematical, real-world oriented nature" (Fayezi et al. 2012: p. 562; Fama and Jensen 1983).

The majority of the selected studies (18 out of 22) adopted agency theory to address problems concerning buyer–supplier relationships. The remaining articles addressed inter-organisational problems (e.g. employer–employee) and other inter-

Table 1.1 Summary of key SCM researches adopting agency theory

Author	Area	Industry	Independent variables	Dependent variables	Stream of agency theory
Agrell et al. (2004)	SCM	Telecom	Asymmetric information, incentive structure	Risk sharing and joint coordination	P-A research
Camuffo et al. (2007)	Strategic management	Air conditioning	Environmental uncertainty, risk monitoring, moral hazard	Risk sharing in vertical inter-firm relations	PAT
Celly and Frazier (1996)	Marketing	Manufacturing	Environmental uncertainty, supplier/distributor characteristics	Coordination efforts	PAT
Cheng and Kam (2008)	SCM	n/a	Structure of network relationships, incentives, supply performance	Dynamics of risk in network collaboration	PAT
Ciliberti et al. (2011)	CSR and SCM	Manufacturing	Code of conduct compliance, moral hazard, adverse selection	Power imbalance in supply chain relationships	PAT
Halldórsson and Skjøtt-Larsen (2006)	Logistics	Fast-moving consumer goods	Goal congruence, risk preference, relationship length	Contract dynamics	PAT
Kudla and Klaas-Wissing (2012)	Procurement	Transportation	Incentive and contractual agreement, adverse selection, moral hazard	Sustainability in supply chain relationships	PAT
Lassar and Kerr (1996)	Strategic management	Electronics	Cost leadership, differentiation and focus	Manufacturer-distributor relationship performance	PAT
Logan (2000)	Logistics	Transportation	Information technologies, norms and values, monitoring, incentive alignment, trust	Outsourcing	PAT
Manatsa and McLaren (2008)	SCM	n/a	Information sharing, incentive alignment	n/a	PAT

(continued)

Table 1.1 (continued)

Author	Area	Industry	Independent variables	Dependent variables	Stream of agency theory
Melek et al. (2015)	New product development	Cross-industry	Asymmetric information, incentive structure	Contractual and relational arrangement of buyer-design firm partnership	PAT
Morgan et al. (2007)	Operations management	Grocery	Supplier opportunism, dependency, monitoring ability, performance outcome, militant behaviour	Category management relationships	PAT
Norrman (2008)	SCM	High-tech	Contract and relational governance, incentive alignment, asymmetric and hidden information, trust	Risk sharing	PAT
Prosman et al. (2016)	Procurement	Cross-industry	Power differences, dependence, goal incongruence, cultural differences, uncertainty	Effectiveness of behavioural-based governance in buyer–supplier relationships	PAT
Ritchie et al. (2008)	SCM	Manufacturing	n/a	Risk, performance	PAT
Simatupang and Sridharan (2002)	SCM	n/a	Mutual objectives, integrated policies, information sharing, incentive alignment	Collaboration	PAT
Tate et al. (2010)	Procurement	Manufacturing	Internal alignment	Relationships, creativity	PAT
Villena et al. (2009)	SCM	Manufacturing	Compensation and employment risk, environmental volatility	Integration	PAT

(continued)

Table 1.1 (continued)

Author	Area	Industry	Independent variables	Dependent variables	Stream of agency theory
Whipple and Roh (2010)	SCM	Manufacturing	Contractual mechanisms	Vulnerability	PAT
Zsidisin and Ellram (2003)	Procurement	n/a	Information asymmetry, programmability of suppliers' activities	Risk management in buyer–supplier relationship	PAT
Zsidisin and Smith (2005)	Procurement	Aerospace	Outcome uncertainty, goal conflict, adverse selection, moral hazard, performance monitoring	Disruption risk in the purchasing organisation–supplier relationship	PAT
Zsidisin et al. (2004)	Procurement	Manufacturing	Outcome uncertainty, goal conflict, adverse selection, moral hazard	Assessment of risk in buyer–supplier relationship	PAT

Fig. 1.1 Journals with agency theory application within supply chain management

organisational situations (e.g. supplier–retailer or supplier–third party). However, considering that in a supply chain multiple actors interact simultaneously, a discrete number of studies (3 out of 22) decided to address supply chain issues taking a triadic and tetradic perspective (Agrell and Norrman 2004; Agrell et al. 2004; Cheng and Kam 2008; Hornibrook 2007). The application of agency theory in networks is useful for researchers and managers aiming to understand how and under what conditions multiple principals and/or multiple agents interact with each other and impact on supply chain behaviours and intentions (Delbufalo and Bastl 2018; Surana et al. 2005).

The application of agency theory in supply chain management contexts follows the indications by Eisenhardt (1989) concerning the "theory-relevant contexts" (p. 70). Specifically, in her 1989 seminal work, Eisenhardt suggests that agency theory can be usefully adopted when:

- principal and agent have different goals (i.e. goal incongruence or goal conflict) such as in buyer–supplier situations (Prosman et al. 2016);
- evaluation of behaviours is difficult due to market or technological uncertainty or task complexity (Morgan et al. 2007; Simatupang and Sridharan 2002; Whipple and Roh 2010);
- there is sufficient outcome uncertainty to increase the performance and relational risks (Fayezi et al. 2012; Melek et al. 2015).

One of the most appealing applications of agency theory in supply chain contexts is in regard to the development of risk mitigation strategies against suppliers' threats of opportunism. Zsidisin and Ellram (2003) address this point by arguing that managing agents' activities using buffers (such as safety stock and multiple sourcing) in conjunction with incentives can be more effective than manipulating agents' behaviour to increase compliance by using, for example, certifications and supplier involvement in target costing development. Zsidisin et al. (2004) and Zsidisin and Smith (2005) corroborate this view by suggesting that the most common principals' strategies to increase compliance (e.g. early supplier involvement) have been used not only to collect information on suppliers' behaviour and mitigate supply chain risks but rather "to lock" suppliers into the supply chain and increase their dependence on the principals (Fayezi et al. 2012: p. 563).

Hornibrook (2007) suggests adopting incentives to moderate the agent's threat of opportunism. He recommends using financial and social incentives in different ways as they have different temporal returns (short term the former and long term the latter). In a similar way, Norrman (2008) asserts that agency relationships tend to be more effective when they are governed through relational and formal (contractual) mechanisms. He also stresses the role of cultural issues and trust in the successful implementation of risk-sharing strategies, suggesting that trust plays a crucial role in increasing the suppliers' compliance intentions and motivations. This is more likely to happen "when risk and reward appear to be equally balanced" (Fayezi et al. 2012: p. 563).

Cheng and Kam (2008) adopt agency theory to evaluate how different supply chain actors respond to unpredictable and unexpected risks. Agency theory assumptions can help actors to analyse the contextual factors responsible for supply chain integration and collaboration (Danese 2011). They suggest that network collaboration is highly linked to several factors such as network structure, supply performance assessment, actors' functional position and mostly the principal–agent contract arrangement (including the definition of incentives, trust building processes and long-term orientation). In particular, the contract arrangement regarding suppliers' performance evaluation processes is of crucial importance in a tiered, multiple supply chain network (Prosman et al. 2016; Ritchie et al. 2008).

A consistent part of the selected articles extensively analyses the supply chain relationship variables, such as incentive alignment, information sharing and reward/sanction development, which can be considered as pillars of agency theory (e.g. Agrell et al. 2004; Manatsa and McLaren 2008; Norrman 2008; Simatupang and Sridharan 2002; Zsidisin and Ellram 2003). Goal incongruence and conflict among supply chain actors are also recurring topics in studies addressing the barriers to supply chain integration and synchronisation (e.g. Morgan et al. 2007; Tate et al. 2010). In the same vein, Melek et al. (2015) investigate inter-organisational integration patterns between the buying firm, design agency and component supplier in NPD projects in order to identify the determinants of these approaches. Asymmetric information, power and incentive structure in supply chain relationships represent relevant topics in the selected literature. For example, Kudla and Klaas-Wissing (2012) analyse dyadic buyer–supplier relationships of logistics services in order to address how the actors send stimuli to each other and respond to such stimuli in conditions of asymmetric information. Prosman et al. (2016) address the power differences among supply chain actors and explore factors influencing the effectiveness of buyer-initiated behavioural-based governance methods.

1.4 Agency Theory Contribution in Supply Chain Management

From the literature review on the contribution of agency theory in the supply chain management literature, a number of considerations can be summarised. Firstly, agency theory provides a useful basis for understanding the multiple aspects of supply chain relationship management: it can explain the impact of a principal's decisions on the agent's propensity to cooperate and comply, as well as to inform contractual responses to (outcome/behavioural) the uncertainty of both actors in the relation (Fayezi et al. 2012). The tenets of agency theory can also contribute to the design of contractual arrangements where mutual information, risk and reward sharing (Cooper et al. 1997; Lee and Whang 2000) and goal congruence (La Londe and Masters 1994) have more chance of being effective, and the establishment and maintenance of long-term business relationships are more easily assured (Cousins 2002).

Secondly, the positivistic agency theory perspective can be usefully applied in the supply chain context in order to understand the potential causes of abnormal behaviours (non-rational or opportunistic) of multiple actors. "Agency theory identifies behavioural change by supply chain actors and sheds light on activities involving principal and agent, self-interest, risk aversion, lack of trust, goal conflict and imperfect policy implementation" (Fayezi et al. 2012: p. 564; Simatupang and Sridharan 2002). Furthermore, the theory suggests how to design contractual arrangements able to moderate or eliminate tensions among supply chain actors by means of information sharing, incentive structure and goal alignment (Knoppen and Christiaanse 2007). Zomorrodi and Fayezi (2011) and Wilson (2006) assert that zones of tolerance and relational elasticity exist in principal–agent relationships. In such situations, behavioural abnormalities are less likely to occur. Such arguments can inform the relationship development in order to reduce the uncertainty typical in supply chain decisions and increase the efficiency of the economic exchange (Prosman et al. 2016).

Finally, agency theory can be adopted to compensate for aspects widely neglected by the supply chain management literature adopting different theoretical perspectives (such as TCE and social exchange theory). As suggested by Fayezi et al. (2012: p. 564), agency theory—mostly from a positivist approach—"extends views centred on task and transaction by attending to specific attributed of agents (or principals) operating within the supply chain". In doing so, agency theory is able to deal with some of the most complex elements of supply chain relationships, such as actors' motivations to cheat, risk preferences and communication. In particular, communication, as an aspect of supply chain relationship development, has received marginal attention in the selected body of literature. This is in contrast to the central role played by communication in orienting actors' behaviours and in contributing to the reduction of supply chain uncertainty.

This literature review—as the others previously published on similar topics (e.g. Fayezi et al. 2012; Bergen et al. 1992)—clearly shows the explanatory power of agency theory in the supply chain management context. However, the adoption of this theory has its limitations. For example, an overemphasis on economic drivers can limit the theory applicability (Heracleous and Lan 2011). According to Fayezi et al. (2012: p. 565), this is in part "an historical legacy, and also occurs because many [supply chain relationships] have traditionally been conceptualised as economic exchange mechanisms, rather than being comprised of complex social and economic relationships".

The agency theory's assumption of "perfect" principal and "imperfect" agent is also questionable and reductive. As Perrow (1986) suggested, the adverse selection and moral hazard agency problems are not restricted solely to the agent sphere of behaviour but also exist on the principal side. This consideration is essential to understand supply chain dynamics where the interdependence and interconnection among the actors are high. In fact, a supply chain is a system where the same actor may act as the principal in a specific relationship and, at the same time, act as the agent in a different one. For instance, a particular distributor who is acting as the principal of a manufacturer (the agent) while at the same time plays the role of agent for the wholesaler (the principal).

Increased organisational complexity—as in the case of an extended network of multiple principals and agents—is another important extension to the agency theory that is useful to improve its contribution to supply chain management literature. The existence of complex arrangements in which multiple actors interact (either multiple principals and/or multiple agents) makes the detection of non-rational behaviours more difficult and challenges the information sharing and monitoring activities. This becomes even more critical when multiple agents have competing interests over a similar (or identical) delegated task, as in the case of code of conduct implementation in a global supply chain discussed later in the study.

References

Agrell, P., & Norrman, A. (2004). Understanding supply chain risk sharing: A three-tier principal-agent approach. In H. Aronsson (Ed.), *16th Anuual Conference for Nordic Researchers in Logistics (NOFOMA)* (pp. 17–33). Linkoping: Linkoping University.

Agrell, P., Lindroth, R., & Norrman, A. (2004). Risk, information and incentives in telecom supply chains. *International Journal of Production Economics, 90*(1), 1–16.

Arrow, K. (1985). The economics of agency. In J. Pratt & R. Zeckhauser (Eds.), *Principals and agents: The structure of business* (pp. 37–51). Boston, MA: Harvard University Press.

Basov, S., & Bardsley, P. (2005). *A general model of coexisting hidden action and hidden information*. Working paper 958, December, Department of Economics, The University of Melbourne, Melbourne.
Bergen, M., Dutta, S., & Walker, O. C., Jr. (1992). Agency relationships in marketing: A review of the implications and applications of agency and related theories. *The Journal of Marketing, 56*(3), 1–24.
Camuffo, A., Furlan, A., & Rettore, E. (2007). Risk sharing in supplier relations: An agency model for the Italian air-conditioning industry. *Strategic Management Journal, 28*(12), 1257–1266.
Celly, K. S., & Frazier, G. L. (1996). Outcome-based and behavior-based coordination efforts in channel relationships. *Journal of Marketing Research, 33*(2), 200–210.
Cheng, S. K., & Kam, B. H. (2008). A conceptual framework for analysing risk in supply networks. *Journal of Enterprise Information Management, 22*(4), 345–360.
Ciliberti, F., de Haan, J., de Groot, J., & Pontrandolfo, P. (2011). CSR codes and the principal-agent problem in supply chains: Four case studies. *Journal of Cleaner Production, 19*, 885–894.
Cooper, M., Lambert, D., & Pagh, J. (1997). Supply chain management: More than a new name for logistics. *The International Journal of Logistics Management, 8*(1), 1–14.
Cousins, P. D. (2002). A conceptual model for managing long-term inter-organisational relationships. *European Journal of Purchasing & Supply Management, 8*(2), 71–82.
Danese, P. (2011). Towards a contingency theory of collaborative planning initiatives in supply networks. *International Journal of Production Research, 49*(4), 1081–1103.
Delbufalo, E. (2012). Outcomes of inter-organizational trust in supply chain relationships: A systematic literature review and a meta-analysis of the empirical evidence. *Supply Chain Management: An International Journal, 17*(4), 377–402.
Delbufalo, E., & Bastl, M. (2018). Multi-principal collaboration and supplier's compliance with codes-of-conduct. *International Journal of Logistics Management* (forthcoming).
Eisenhardt, K. M. (1989). Agency theory: An assessment and review. *Academy of Management Review, 14*(1), 57–74.
Fama, E. F. & Jensen, M. C. (1983). Separation of ownership and control. In J. B. Barney & W. G. Ouchi (Eds.), *Organizational economics*. California: Jossey-Bass Inc. Publishers (Reprinted in slightly adapted form from the *Journal of Law and Economics 26*).
Fayezi, S., O'Loughlin, A., & Zutshi, A. (2012). Agency theory and supply chain management: A structured literature review. *Supply Chain Management: An International Journal, 17*(5), 556–570.
Fleisher, C. S. (1991). Using an agency-based approach to analyze collaborative federated interorganizational relationships. *The Journal of Applied Behavioral Science, 27*(1), 116–130.
Halldórsson, A., & Skjøtt-Larsen, T. (2006). Dynamics of rela-tionship governance in TPL arrangements: A dyadic perspective. *In-ternational Journal of Physical Distribution & Logistics Management, 36*(7), 490–506.
Harris, M., & Raviv, A. (1979). Optimal incentive contracts with imperfect information. *Journal of Economic Theory, 20*(April), 231–259.
Hart, O., & Holmström, B. (1987). The theory of contract. In T. Bewley (Ed.), *Ad-vanced in economic theory, Fifth world congress*. London: Cambridge University Press.
Heracleous, L., & Lan, L. L. (2011). Agency theory, institutional sensitivity, and inductive reasoning: Towards a legal perspective. *Journal of Management Studies, 49*(1), 223–239.
Hornibrook, S. (2007). *Agency theory and supply chain management: Goals and incentives in supply chain organisations*. Working paper 147, 24 July, Kent Business School, University of Kent, Canterbury.
Jensen, M. C., & Meckling, W. H. (1976). Theory of the firm: Managerial behavior, agency costs, and ownership structure. *Journal of Financial Economics, 3*(4), 305–360.
Knoppen, D., & Christiaanse, E. (2007). Supply chain partnering: A temporal multidisciplinary approach. *Supply Chain Management: An International Journal, 12*(2), 164–171.

Kudla, N. L., & Klaas-Wissing, T. (2012). Sustainability in shipper-logistics service provider relationships: A tentative taxonomy based on agency theory and stimulus-response analysis. *Journal of Purchasing & Supply Management, 18,* 218–231.

La Londe, B., & Masters, J. (1994). Emerging logistics strategies: Blueprints for the next century. *International Journal of Physical Distribution & Logistics Management, 24*(7), 35–47.

Lassar, W. M., & Kerr, J. L. (1996). Strategy and control in supplier-distributor relationships: An agency perspective. *Strategic Management Journal, 17*(8), 613–632.

Lee, H., & Whang, S. (2000). Information sharing in a supply chain. *International Journal of Manufacturing Technology and Management, 1*(1), 79–93.

Logan, M. S. (2000). Using agency theory to design successful outsourcing relationships. *The International Journal of Logistics Management, 11*(2), 21–32.

Manatsa, P. R., & McLaren, T. S. (2008). Information sharing in a supply chain: Using agency theory to guide the design of incentives. *Supply Chain Forum: International Journal, 9*(1), 18–26.

Melek, A. A., Van den Ende, J., & Ianniello, G. (2015). Inter-organizational coordination patterns in buyer-supplier-design agency triads in NPD projects. *International Journal of Operations & Production Management, 35*(11), 1512–1545.

Mitnick, B. M. (1973). Fiduciary rationality and public policy: The theory of agency and some consequences. 1973 Annual Meeting of the American Political Science Association, New Orleans, LA. In *Proceedings of the American Political Science Association.*

Morgan, N. A., Kaleka, A., & Gooner, R. A. (2007). Focal supplier opportunism in supermarket retailer category management. *Journal of Operations Management, 25*(2), 512–527.

Norrman, A. (2008). Supply chain risk-sharing contracts from a buyers' perspective: Content and experiences. *International Journal of Procurement Management, 1*(4), 371–393.

O'Loughlin, A., & Clements, M. (2007). Relational exchange, supply chain myopia and zones of turbulence. In Sohal, A. (Ed.), *Proceedings of the Eighth International Research Conference on Quality Innovation and Knowledge Management (QIK),* New Delhi (pp. 610–620).

Perrow, C. (1986). *Complex organizations: A critical essay* (3rd ed.). New York: Random House.

Prosman, E. J., Scholten, K., & Power, D. (2016). Dealing with defaulting suppliers using behavioral based governance methods: An agency theory perspective. *Supply Chain Management: An International Journal, 21*(4), 499–511.

Ritchie, B., Brindley, C. S., & Armstrong, N. (2008). Risk assessment and relationship management: Practical approach to supply chain risk management. *International Journal of Agile Systems and Management, 3*(3/4), 228–247.

Ross, S. A. (1973). The economic theory of agency: The principal's problem. *The American Economic Review, 63*(2), 134–139.

Shapiro, S. P. (2005). Agency theory. *Annual Review of Sociology, 31,* 263–284.

Simatupang, T., & Sridharan, R. (2002). The collaborative supply chain. *The International Journal of Logistics Management, 13*(1), 15–30.

Stock, J. (1997). Applying theories from other disciplines to logistics. *International Journal of Physical Distribution & Logistics Management, 27*(9/10), 515–539.

Surana, A., Kumara, S., Greaves, M., & Raghavan, U. (2005). Supply-chain networks: A complex adaptive systems perspective. *International Journal of Production Research, 43*(20), 4235–4265.

Tate, W. L., Ellram, L. M., Bals, L., Hartmann, E., & Van Der Valk, W. (2010). An agency theory perspective on the purchase of marketing services. *Industrial Marketing Management, 39*(5), 806–819.

Villena, V. H., Gomez-Mejia, L. R., & Revilla, E. (2009). The decision of the supply chain executive to support or impede supply chain integration: A multidisciplinary behavioral agency perspective. *Decision Sciences, 40*(4), 635–665.

Whipple, J. M., & Roh, J. (2010). Agency theory and quality fade in buyer-supplier relationships. *The International Journal of Logistics Management, 21*(3), 338–352.

Williamson, O. E. (1988). Corporate finance and corporate governance. *The Journal of Finance, 43*(3), 567–591.

Wilson, M. (2006). *Supply chains behaving badly: A dynamic model of inter-organisational supply chain exchange behaviour under rational, relational and chaotic paradigms (Ph.D.)*. Lincoln University, Lincoln.

Zomorrodi, M., & Fayezi, S. (2011). Understanding the concept of elasticity in supply chain relationships: An agency theory perspective. *Asian Journal of Management Research, 1*(2), 452–472.

Zsidisin, G. A., & Ellram, L. M. (2003). An agency theory investigation of supply risk management. *Journal of Supply Chain Management, 39*(3), 15–27.

Zsidisin, G. A., & Smith, M. E. (2005). Managing supply risk with early supplier involvement: A case study and research propositions. *Journal of Supply Chain Management, 41*(4), 44–57.

Zsidisin, G. A., Ellram, L. M., Carter, J. R., & Cavianto, J. L. (2004). An analysis of supply risk assessment techniques. *International Journal of Physical Distribution & Logistics Management, 34*(5), 397–413.

Chapter 2
Extending Sustainability to Suppliers: The Role of Codes of Conduct

Abstract The chapter addresses the antecedents of suppliers' (mis)conduct along with the role and function of the buyers' code of conduct in global supply chains. The chapter also presents the main characteristics of multi-stakeholder collaborative initiatives developed to respond to multiple stimuli coming from customers and society. Specific attention is dedicated to the difficulties experienced by MNCs in monitoring supplier's compliance in complex environments.

Keywords Compliance · Code of conduct · Collaboration

2.1 Antecedents of Suppliers' Conduct

The widely known definition of sustainability provided by the World Commission on Environment and Development (1987: p. 8)—"development that meets the needs of the present without compromising the ability of future generations to meet their needs"—recalls the concept of triple bottom line (Elkington 1998) which simultaneously considers economic, environmental and social issues in managing global supply chains. Despite the growing interest in this topic by academics and scholars (for more details, see Aguinis and Glavas 2012), recently there have been rising concerns about the effective applicability of the concept to actual supply chain strategies. One of the most challenging aspects is that the boundary of responsibility extends beyond the reach of a corporation's ownership and control (Gimenez and Tachizawa 2012). A high level of social and environmental performance depends on the corporation's activities and—in a more decisive way—on the decisions and performance of a multiplicity of other supply chain actors (mostly suppliers) operating independently and often in different geographical contexts (Amaeshi et al. 2008).

Despite many efforts by MNCs to implement social and environmental issues in their supply chains, a gap exists between the intention of supply chain sustainable management and its implementation in practice. The extension of suppliers' decep-

E. Delbufalo, *Agency Theory and Sustainability in the Global Supply Chain*, SpringerBriefs in Business, https://doi.org/10.1007/978-3-319-72793-6_2

tion is still high,[1] thus limiting the potential benefits linked to sustainability policies (Bowen et al. 2001). Ross and Chan (2002), Jiang (2009a) recall that some suppliers in developing countries believe that higher labour costs and various environmental standards are examples of disguised protectionism designed to keep their exports out of developed countries' markets. Even if business ethics and social–environmental standards are widely established in developing countries, they are "set in stone" and are not effectively enforced or fully understood by some suppliers (Jiang 2009b: p. 268). Roberts and Engardio (2006) corroborate such a view by showing that in China there are a discrete number of consultants who assist firms in evading buyers' audits and standards. It is also estimated that only 20% of Chinese suppliers comply with wage rules, and only 5% obey working hour limitations.

Considering the high degree of suppliers' deception, the most urgent question becomes the following: What are the antecedents of suppliers' (mis)conduct? Recent literature proposes three main factors that are able to influence suppliers' behaviour for sustainability purposes (e.g. Krause et al. 2007; Moran 2005): price pressure, production complexity and contract duration. While cost brings price pressure to suppliers, the others represent the pressure of production complexity on suppliers (McEvily and Marcus 2005). Jiang (2009b: p. 268) efficaciously summarises the pressure the suppliers are facing: "The globalized world economy has formed a pyramid-like, multi-layer system of contracts […] MNCs are on top of the pyramid and hold the ultimate right to command; suppliers in developing countries usually cluster in the bottom, marked by low-skilled work and low-tech competence […]. The buyer-dominated atmosphere in this pyramid-like system of contracts may lead to unachievable targets set by buyers and unfair distribution of responsibility levied on the suppliers".

Price pressure: one of the main rationales beyond MNCs' sourcing from developing countries is the search for cost savings (Short et al. 2015). In many cases, suppliers tend to respond with unethical and questionable actions to the pressure of the buyers for price minimisation. As a consequence, suppliers often pass down the price pressure to workers by reducing workers' rights and compressing the investments for improving working conditions and safety (Jiang 2009b). Anecdotal evidence also shows how suppliers often decide to use less costly but non-authorised materials or components during production to deal with the price pressure (for details, see Fairclough 2007).

Production complexity: a commonly known list of buyers' requirements for suppliers includes high-quality products, timely delivery, smaller production runs at more frequent intervals and a high degree of responsiveness towards changes in vol-

[1] For example, in 2007 Mattel had to recall 20 million children's toys because a significant part of their suppliers used toxic materials. A Clean Clothes Campaign (2016) Report shows that, after a ten-year-long period of commitment to worker safety, the major apparel brands and retailers that are part of the Alliance for Bangladesh Worker Safety—including Gap Inc., Target, VF Corporation, Hudson's Bay Company and Walmart—are still struggling in their supply network with unsafe factories and hundreds of thousands of workers at risk. Large global brands (such as Adidas and Nike) experienced similar problems with South Asian manufacturers of footballs with regard to child labour practices (Lund-Thomsen and Nadvi 2011).

ume (e.g. Dyer 1996). As buyers increase the just-in-time logic for production and delivery, the timing left to the suppliers to react to changes is progressively shortened. Thus, "to deal with short lead time and high flexibility, suppliers in developing countries may either force excessive overtime or speed up production lines to workers' maximum physical limits" (Jiang 2009b: p. 268).

Contract duration: a growing number of studies on supply chain coordination issues show that suppliers' performance is highly correlated to long-term relationship decisions. In fact, suppliers are more willing to make resource investments—finalised to reduce the social and environmental impact of their activities—when buyers show the intention to invest in long-term partnerships (e.g. Cerruti et al. 2016). In long-term relationships, it is also plausible that buyers and suppliers would be willing to reciprocally make relation-specific investments—investments often too costly for the suppliers to sustain autonomously—dedicated to improving supply chain integration and reducing the gap between actual performance and buyers' sustainability targets (e.g. Krause et al. 2007).

While the above-mentioned arguments contribute to explain the rationale behind suppliers' motivations for defection (or limited compliance with social and environmental standards), they can be sufficient to predict a supplier's conduct during economic exchanges. Price pressure, production complexity and contract duration can be considered as antecedents of a supplier's conduct, but they do not contribute directly to negative consequences; we need a trigger event to explain how such sources of complexity stimulate opportunistic behaviour by non-ethical suppliers. Besides, since the same factor (i.e. antecedent) may lead to different consequences, we need to analyse the element most responsible for determining suppliers' behaviour: the governance mechanism adopted in a buyer–supplier relationship (Delbufalo 2012).

A wide body of research in supply chain management demonstrates that "high levels of commitment and low levels of opportunism characterize effectively governed interorganizational relationships between financially independent firms" (Jiang 2009b: p. 269). Commitment refers to the willingness of interacting partners to comply with the negotiated practice on behalf of the transaction (Fynes and Voss 2002). Opportunism "refers to a lack of candor or honesty in transactions, to include self-interest seeking with guile" (Williamson 1985: p. 9). It includes withholding or misrepresenting information to "mislead, distort, obfuscate, or otherwise confuse" (Williamson 1985: p. 47). It can also be cheating, "not delivering the promised action […] and failing to do this on a fairly systematic and sustained basis" (Hardy and Magrath 1989: p. 123). According to these arguments, violations against the pre-agreed standards (often included in code of conduct or other CSR documents) can be seen as suppliers' opportunistic behaviour (Moore 2001).

Agency theory—as described in Chap. 1—suggests that every actor in economic exchange is assumed to be affected by bounded rationality, risk aversion and opportunistic tendency (Eisenhardt 1989), but opportunism in interorganisational relations may be controlled through threats (as in market governance) or incentives (as in hierarchical governance) (Gundlach et al. 1995; Jiang 2009a). Two different governance mechanisms can be developed in the CSR context, when the buyer (often an MNC) enters into a relationship with one or more suppliers operating in developing countries and needs to manage the relationship, assuring sustainability within the supply process. The first mechanism is the external market pressures (alternatively named as vertical governance in buyer–supplier relationships). Here, the MNCs enforces the suppliers to adopt social and environmental policies—reflecting their corporate culture, values and traditions—and ask the suppliers to accept the increased costs linked to monitoring and improving investments. If a supplier is found to be consistently and repeatedly non-compliant, MNCs may decide to terminate the supply contract. As an alternative to such vertical governance, MNCs may decide to operate on a socially mutual adaptation register (Gundlach et al. 1995) and make idiosyncratic investments (e.g. *ex-ante* and *ex-post* training and incentives) to participate in the costs and efforts required to the suppliers. "Buyer and supplier concerns of the safety of the idiosyncratic investments contribute to the development of mutual safeguards" (Kostova and Zaheer 1999; Jiang 2009b: p. 270). In such a situation, the responsibilities of buyers and suppliers are strictly interconnected and mutually reinforcing. This second—and more participative—governance mechanism can be defined as horizontal or peer-to-peer governance. In the following sections, we will discuss the implications of the choice between the two governance approaches and also discuss how these antecedents affect the consequences (i.e. compliance with social and environmental standards).

2.2 Corporate Code of Conduct in Global Supply Chains

A code of conduct is a formal CRS document employed by companies (often MNCs) "to establish and communicate responsible business practices and an ethical organizational culture" (Erwin 2011). These institutional documents, also referred to as a "codes of ethics" or "codes of business standards", are developed to detail in an explicit way a corporation's commitment to CSR and the expected conduct from the organisation's employees and partners. According to Collins (2004), the code of conduct is a fundamental tool to formalise social and environmental targets of industrial and commercial activities within an organisation as well as to create and sustain an ethical organisational culture and tradition. "Ideally, codes of conduct affect the organizational culture by governing the actions and conduct of employees through the promotion of ethical business practices, thereby avoiding legal consequences. Further, as tangible commitments to social responsibility, adopting codes of conduct may lead to reputational benefits by functioning as a symbol of CSR awareness and engagement, thereby preserving and legitimating the company's public image" (Erwin 2011: p. 536; Diller 1999). According to major CSR literature, the development of a code of conduct is correlated to several other benefits, such as reduced negative consumer actions, product differentiation and better customer relations (Diller 1999; Lenox and Nash 2003).

However, the positive return of the code's adoption is strongly linked to its quality, which is recognised as a critical component of code effectiveness. In fact, factors such as comprehensiveness, clarity and scope determine the ability of the code to be understood by its multiple users and influence the appropriate monitoring of measures and standards. Erwin (2011: p. 536) reflects on this matter saying: "As the adoption of corporate codes becomes the norm, the incidence of insincere, rhetoric-based codes never designed to move beyond a cursory treatment of issues (i.e., greenwashing) is likely to increase. Analysis of code effectiveness based on code presence and absence may group companies with perfunctory codes alongside top CSR performers and obscure empirical results. Thus, a natural progression of CSR research is to investigate the importance of the quality, or comprehensiveness, of these codes and their effect on CSR performance".

To design an effective code of conduct is particularly important if we consider that the code of conduct contributes to the development of an ethical organisational culture, which spreads across the boundaries of the single firm (promoting the initiative) and engages all the other actors in the supply chain. In fact, since the beginning (during the 1990s), MNCs started to acknowledge—also developing appropriated measures and standards to be monitored—a responsibility for working conditions at their suppliers' factories in developing countries. Here, "the codes of conduct emerged as the dominant way for MNCs to operationalise this extended sense of responsibility" (Egels-Zandén 2007: p. 45; Radin 2004; Sethi 2002).

From a supply chain perspective, codes are instruments that facilitate communication between parties in a network that have both direct and (mostly) indirect relationships. Through the codes' provisions, manufacturers can inform final customers, NGOs or society about certain features of their products and the social and environmental performance of their processes. These features might not be immediately visible for the other parties so that a public statement is necessary to confirm, for example, that no child labour has been encouraged or no polluting emissions during the production process have been produced. What is formalised in the code becomes, in most cases, more credible to the customers and the other stakeholders as the MNCs (promoting the code) take responsibility for the formal statements. Furthermore, if a third independent party monitors—and certifies—whether the producer lives up to these claims and publishes publicly about the results, this will enhance credibility, thus increasing the utility of the code itself (Ciliberti et al. 2009).

Recently, compliance with codes of conduct has been included as a substantial requisite in contracts between supply chain partners. This inclusion aims to reduce contractual incompleteness and improve supply chain coordination. In a supply chain contract are also detailed the procedures and standards for the code's monitoring, which include auditing processes and data collecting. Such activities reduce information asymmetry among supply chain partners because the way partners execute processes becomes more transparent and the information sharing more reliable.

2.2.1 Limits of the Implementation of Code of Conduct in Emerging Countries

In spite of MNCs' commitment and the undeniable benefits of codes of conduct development in supply chain contexts, several factors have been pervading their effective implementation (Delbufalo and Bastl 2018).

Some of them refer to the code's characteristics and the way in which costs and benefits are distributed among the actors involved. Many studies suggest that a limited uniformity across codes makes it difficult for the suppliers to comply with the provisions required by many buyers at the same time (Risso 2012). In addition, Emmelhainz and Adams (1999) suggest that a large portion of codes of conduct contains inconsistent details in the areas of monitoring and enforcement, which makes it difficult to define the auditing processes and analyse the results. The implementation of codes can be costly and time-consuming, and the related potential benefits might not be clearly identified or be unevenly distributed among supply chain partners (Pedersen and Andersen 2006). This is even more important when SMEs are involved in the transactions, as price pressure and production complexity compress in a more consistent way—sometimes making it almost economically impossible to comply with the agreement provisions—the resources available for investments and increase the motivations for defection (for a detailed analysis on SMEs difficulties in complying with CSR initiatives, see Ciliberti et al. 2011).

The effective implementation of codes of conduct can be strongly limited by the so-called decoupling between policy and practice (Egels-Zandén 2014). In relation to codes of conduct, the concept of decoupling implies that suppliers may be more interested in obtaining the legitimacy and signalling the benefits of perceived code of conduct compliance than in fully implementing the practices prescribed by the buyers (e.g. DiMaggio and Powell 1983). This is so because, for example, improving workers' rights and welfare (including correct overtime compensation, instituting salary increases and limits on overtime) generally increases production costs and lowers suppliers' ability to meet buyers' strict production deadlines (Egels-Zandén 2014). Anecdotal evidence confirms that suppliers are more interested in performing well during the monitoring activities (also obtaining high scores in auditing) than in actually improving social and environmental impacts in practice. "This implies that code of conduct compliance risks becoming a façade of conformity that suppliers erect because buyers demand it. If successful, this façade satisfies buyers, allowing suppliers to preserve internal arrangements that violate workers' rights without sacrificing access to external resources" (Egels-Zandén 2014: p. 61).

Difficulties in implementing codes of conduct may also refer to the degree of relationship integration and commitment between buyer and supplier. Here, one of the most critical factors is the risk of opportunism and information appropriation, as MNCs could utilise suppliers' information—gathered for monitoring of suppliers' social and environmental performance—for increasing their control over the supply base (Fernandez and Chiambaretto 2016). The difference between a firm's need to share information to ensure the success of a common project and its need to limit information sharing to avoid informational spillovers between the collaborating partners generates tensions among buyers and suppliers and challenges the effectiveness of code of conduct actual adoption (Tidström 2014). Furthermore, suppliers' degree of commitment to the adoption of codes of conduct and other similar initiatives are not uniform. In fact, anecdotal evidence suggests it may vary significantly, as some suppliers welcome collaborative actions and monitoring, while others feel uncomfortable with them, at least initially. As stated in the work of Plambeck et al. (2012) "The experience of having a group of auditors simultaneously poring over a factory's records can be intimidating for suppliers. As a result, it pays to start off [buyer-supplier collaboration] slowly to accustom suppliers to the concept and see its benefits. This may mean simply sharing findings about a factory in the beginning rather than immediately conducting joint on-site audits" (Plambeck et al. 2012: p. 48). Such varying degrees of commitment can introduce tensions in the relationship between the buyer and an uncommitted (or less committed) supplier, which requires a carefully thought through approach to relationship management and roll-out of CSR initiatives.

Finally, a number of other factors have been identified by CSR literature to explain the difficulties in the implementation of codes of conduct in developing countries, among which the most often cited are the following:

- *Buyer characteristics*: Lund-Thomsen and Nadvi (2011) compared football production clusters in Sialkot, Pakistan and Jalandhar (India) and find that codes of conduct have had greater effects in the country where the buyers are more branded. Frundt (2000), Armbruster-Sandoval (2005) support this argument by suggesting that codes of conduct are most effective when the buyer has a well-known brand name. Additional evidence comes from Barrientos and Smith (2007); they show codes of conduct have greater impact when the suppliers see many of their buyers imposing similar demands. "The underlying argument here is that branded companies impose more stringent external demands and surveillance that lead to tight couplings of policy and practice" (Egels-Zandén 2014: p. 63; Hallett 2010; Kelly and Dobbin 1998).
- *Supplier characteristics*: Locke et al. (2007) assert code of conduct compliance is positively related to suppliers' foreign (vs. domestic) ownership. Koçer and Fransen's (2009) study partly supports this finding, demonstrating that compliance is positively related to Western ownership of the supplier, whereas limited improvement is to be expected with other types of ownership. Adding to this, Locke et al. (2007) provide empirical evidence that compliance depends on supplier size; they find smaller factories generally to be more compliant than larger factories. Egels-Zandén (2014) link these results to the more general new institutional literature, which argues that tight couplings are more likely in organisations with specific characteristics (see also Clemens et al. 2008).
- *Governance mechanism of buyer–supplier relationships*: Lim and Phillips (2008) conducted a study of Nike's Chinese and Vietnamese suppliers in the early 2000s and found that codes of conduct could improve workers' rights if buyer–supplier relationships were transformed from arm's length to collaborative relationships.

This finding can be linked to new institutional theory suggesting how the type of relationship between interacting partners can influence the likelihood of recoupling (Frenkel and Scott 2002).

- *Supplier's cultural attitudes* towards ethical issues and transparency policy (e.g. Jiang 2009a; Lund-Thomsen 2008; Taylor 2011).
- *Limited/intermittent monitoring* or a "ceremonialisation" of the auditing practices (Esbenshade 2004; Hill and Rapp 2014).

2.2.2 *The Monitoring of Suppliers' Compliance*

Monitoring is a crucial activity of the code of conduct implementation process. In CSR initiatives, formal monitoring has two main objectives: first, it enables the buyer to signal to all relevant stakeholders (*in primis* suppliers, customers and shareholders) the intention to commit to ethical, social and environmental issues and, during the process, to legitimate the corporation's CSR efforts. Second, "monitoring represents an attempt by the buyer to ensure supplier adherence to CSR guidelines and gauge progress in CSR implementation" (Boyd et al. 2007: p. 344).

MNCs have shown divergent approaches in monitoring suppliers' behaviour. The high variability in terms of frequency, depth and breadth of the process makes it difficult for consumers, shareholders and other stakeholders to compare the performances of different supply chains. Both researchers and practitioners also claim that the lack of uniformity of monitoring has led to suppliers questioning the seriousness and commitment of the buyers towards CSR issues. Michael Kobori of Levi Strauss, for example, recently identified increasing the uniformity and effectiveness of supplier monitoring as a top priority for supply chain managers looking to the future (Kondej 2016).

The rationale behind greater efforts towards supplier monitoring of codes of conduct is based on the assumption that higher levels of monitoring will increase the likelihood of supplier compliance by reducing information asymmetry in buyer–supplier relationships. A reduction in information asymmetry implies that the suppliers will not be able to hide non-compliant behaviour and will be encouraged to adhere to the buyers' requests. However, the literature suggests the increase in the level of monitoring is beneficial only until a specific decoupling point, beyond which a higher increase may not be the best strategy for increasing supplier compliance. For example, Murry and Heide's (1998) study of retailer compliance with manufacturer initiatives within distribution channels finds a negative relationship between manufacturer monitoring of retailers and retailers' compliance with established promotional agreements with manufacturers. They argue that high levels of monitoring can signal distrust on the part of the monitoring party, and note that a common reaction by the distrusted party is retaliatory behaviour such as non-compliance (Boyd et al. 2007). A perceived excessive monitoring also increases the tendency of the suppliers—who have been forced to disclose their activities to the buyers—to push the boundaries and engage in non-productive (even harmful) behaviour just to show aversion to the buyers' actions.

Furthermore, an invasive monitoring can also have negative implications for the relational dimensions of supply chain relationships, such as trust and commitment. The suppliers' feeling that they are being over-monitored can provoke conflicts as they sense a loss in their strategic and operational autonomy. Such situations will tend to undermine the development of high levels of trust and commitment, which prior research suggests as being essential to successful joint actions and information sharing in exchange relationships (Boyd et al. 2007; Tomkins 2001).

There are different manners for MNCs to monitor the actual compliance with the code of conduct by their suppliers. The monitoring process can be managed by multiple actors with different responsibilities and functions. First, national or local government may provide a general control over the social and environmental performance of supply chain actors. For example, government labour ministries or specialised government agencies oversee workplace standards. This system remains the norm in the developed countries, while few developing countries have an effective government system to enforce labour standards and other ethical requirements. In light of these inadequacies, MNCs managing supply networks in developing countries are loath to accept local enforcement as sufficient, much less the mere existence of formal legal standards.

Second, to enrich the control over the code of conduct's standards, many MNCs build a self-monitoring system. Buyers can gather information on a supplier's compliance by a third party or by themselves (by using internal auditors). Third parties are usually specialist companies, or institutions, possessing unique competences related to a specific geographical or institutional context, which buyers do not necessarily possess. A buyer's decision to include third parties in CSR initiatives is primarily driven by the need to reduce ambiguity and risk of misinterpretation of the data collected. In case of high ambiguity, buyers run the risk of not being able to recognise (and prove) that suppliers have not complied with the agreement (Milgrom and Roberts 1992). In such a situation, the role of a third party serves as a protective mechanism in reducing the code's ambiguity and preventing violations of the code of conduct through a correct interpretation of the policies and standards. Furthermore, buyers may use third parties for ensuring transparency and objectivity in the information acquisition and elaboration, further increasing the legitimacy of the process. In fact, customers and other stakeholders might perceive third-party monitoring as being more reliable and credible, thus contributing to the positive return of CSR investments (Pedersen and Andersen 2006). While the above-mentioned reasoning argues in favour of the inclusion of third parties in code of conduct implementation, some studies question the value of doing so. O'Rourke (2002) showed that a third-party's failure to accurately assess suppliers' compliance with codes of conduct decreases the buyer's ability to make fully informed strategic decisions and subjects these firms to the risk of negative reputational consequences. For example, in a recent work of nearly 17,000 supplier audits, Short et al. (2015) revealed that auditors report fewer violations in specific and detailed circumstances (e.g. when individual auditors have audited the factory before, when audit teams are less experienced or less trained and when audits are paid for by the audited supplier). Here, failure of the third party to provide the buyers with accurate and relevant information about suppliers' compliance with codes of conduct has the opposite of the desired effect and increases information asymmetry between the buyer and the potentially non-compliant supplier. Alternatively to the inclusion of a third party (i.e. external auditing) is the choice of internal auditing. This happens when MNCs set up their own independent systems for monitoring conditions in their plants and those of their suppliers. These systems receive corporate funding but have their own independent management. Mattel has used this approach, for example, in establishing the Mattel Independent Monitoring Council (MIMCO), which conducts factory audits under the guidance of independent experts (Bremer and Udovich 2001). Both of these approaches (external and internal monitoring) are likely to be relatively expensive, given the potential requirement to monitor hundreds, or even thousands, of supplier firms spread across various regions and countries. Often, considering the high costs of monitoring, the buyers decide to limit the monitoring to some selected suppliers, most closely tied to the buyers. In doing so, they neglect the performance of the suppliers at the bottom of the supply chain. "Since the latter are generally regarded by industry experts to be the worst offenders, this approach may not be effective in curbing serious abuses, even if it is successful in the near term in meeting public relations objectives" (Bremer and Udovich 2001: p. 337).

Third, the MNCs may decide to share the monitoring activity's costs (and benefits) with some other corporations and, for such purpose, form a so-called monitoring coalition (Delbufalo and Bastl 2018; Risso 2012). The aim of the coalition is twofold: firstly, to expand and enrich the informative range of monitoring by collaborating with other stakeholders (firms, NGOs, etc.) and, secondly, to mutually benefit from other partners' experience in defining a shared code of conduct and action plans for implementation. The following section describes, in more detail, the benefits and challenges of different multi-stakeholder monitoring coalitions.

2.3 Collaboration and Multi-stakeholders' Code of Conduct

In recent times, in the social and environmental policy domain, national regulation has been gradually complemented by standards issued by complex multi-stakeholder initiatives involving MNCs, international organisations, NGOs and public agencies (Cashore et al. 2007; Mele and Schepers 2013; Porter and Ronit 2006; Prakash and Potoski 2007; Radaelli 2003; Stafford 2007). The reason behind multi-stakeholder collaboration for sustainability is that a single firm initiative (and a corresponding corporate code of conduct) often lacks legitimacy from society and firms' partners, thus becoming an ineffective instrument for collecting information about social and ethical issues (Böstrom 2006).

Multi-stakeholders' codes of conduct are created "when NGOs, multi-lateral, and other organizations encourage companies to participate in schemes that range from setting social and environmental standards to ensuring some forms of monitoring and certification. [...] The (minimum) common denominator of these schemes is the stated goal of encouraging stakeholder dialogue and social learning, thus enhancing the legitimacy of the participants" (Mele and Schepers 2013: p. 562; Baccaro and Mele 2011). The peculiarity of such initiatives is that MNCs are participants rather than initiators or principals—and thus need to commit to the process development—but "are only partly in control of the process" (Roloff 2007: p. 234). The outcome of multi-stakeholders' initiatives (code of conduct or guidelines) is substantially different from industry self-regulation, in which private sector organisations unilaterally develop best practices and technical standards along with an implementation process, including "internal management system, auditing and accounting by third parties" (Haufler 2003: p. 238).

In contrast, a multi-stakeholder collaboration may benefit from a comprehensive membership base composed of a variety of different actors from the public, private and civil society engaged in developing (through a negotiation process) a regulatory—although voluntary—set of initiatives and tools (Fransen and Kolk 2007). The presence of different actors with complementary requests and needs can be considered the most important enabler of increased legitimacy. Such membership variety increases credibility and authority of the initiative since "the scheme conveys an image of neutrality and its stable institutional arrangement is thought to enable cooperation, dialogue, and interdependence among participants. At the same time, broad representation increases the likelihood of information asymmetries and power struggles, thus threatening the very existence of these initiatives" (Mele and Schepers 2013: p. 563).

A wide body of the literature suggests that a multi-stakeholder collaboration contributes to the institutionalisation of CSR practices extending beyond the boundaries of a single initiative (Scott 2008). The way in which multiple and different actors collaborate over the long term contributes to the development of an institutional infrastructure dedicated to sustainable practices: in doing so, the collaborating partners establish a set of norms and procedures, reframing what companies need to do to be legitimately recognised as socially responsible actors by customers, suppliers and other relevant parties (Fransen and Kolk 2007).

However, a few contrasting arguments come from Schäferhoff et al. (2009), Bernstein (2005) and Bäckstrand (2006). They claim that, in many cases, a multi-stakeholder initiative results in a complex institutional arrangement having as its main purpose the continuous search for equilibrium between coopetition and mutual dependency among participants. In fact, the interaction among different actors—such as competitors in the same industry and/or labour unions—can give rise to several tensions linked to, for example, information sharing and privacy protection. The definition of a multi-stakeholders' code of conduct—with the corresponding set of monitoring and results disseminating activities—can increase the MNCs' concerns about the use of competitive information to obtain unethical advantages. All the participants can use shared information and knowledge to analyse environmental and social deficiencies; they can learn from one another and jointly develop corrective actions that are mutually beneficial (Arya and Salk 2006). However, most of such participants have contrasting needs, such as in the case of competitors in the same industry; they have to protect their core competencies from the other actors, because companies operating in the same industry must develop unique skills (Nelson and Winter 1982). Information that is shared between the competing actors about a common collaborative project could be potentially misused, for purposes of obtaining individual companies' benefits and advantages, against the other members. The dilemma between an actor's need to share information in order to ensure success of the collaborative initiative and the need to limit information sharing to protect its sources of competitive advantage generates tensions and reduces the potential benefits of collaborative initiatives (Arya and Salk 2006).

A growing consensus exists regarding the following statement: the effectiveness of the initiative depends on the participants' attitude and intentions towards the collaborative efforts. In other words, it depends on "whether the corporation interprets the multi-stakeholder initiative as a genuine opportunity to engage in dialogue with and learn from other organizations, or merely as pure window-dressing" (Mele and Schepers 2013: p. 563). Following this reasoning, during the process of institutionalisation of the multi-stakeholders initiative, the participants need to carefully develop appropriate governance mechanisms to manage the tensions linked to divergent interests. Control mechanisms are essential to facilitate interactions between the stakeholders while increasing the incentives for each participant to eliminate (or at least reduce) opportunistic behaviours. In doing so, they foster the success of collaborative efforts while limiting the negative competitive repercussions and the resulting conflicts (Das and Teng 1998).

References

Aguinis, H., & Glavas, A. (2012). What we know and don't know about Corporate Social Responsibility: A review and research agenda. *Journal of Management, 38*(4), 932–968.

Amaeshi, K. M., Osuji, O. K., & Nnodim, P. (2008). Corporate social responsibility in supply chains of global brands: A boundaryless responsibility? Clarifications, exceptions and implications. *Journal of Business Ethics, 81*(1), 223–234.

Armbruster-Sandoval, R. (2005). *Globalization and cross-border labor solidarity in the Americas: The anti-sweatshop movement and the struggle for social justice*. New York: Routledge.

Arya, B., & Salk, J. (2006). Cross-sector alliance learning and effectiveness of voluntary codes of corporate social responsibility. *Business Ethics Quarterly, 16*(2), 211–234.

Baccaro, L., & Mele, V. (2011). For lack of anything better? International organizations and global corporate codes. *Public Administration, 89*(2), 451–470.

Bäckstrand, K. (2006). Multi-stakeholder partnerships for sustainable development: Rethinking legitimacy, accountability and effectiveness. *European Environment, 16,* 290–306.

Barrientos, S., & Smith, S. (2007). Do workers benefit from ethical trade? Assessing codes of labour practice in global production systems. *Third World Quarterly, 28*(4), 713–729.

Bernstein, S. (2005). Legitimacy in global environmental governance. *Journal of International Law and International Relations, 1*(1/2), 139–166.

Böstrom, M. (2006). Regulatory credibility and authority through inclusiveness: Standardization organizations in cases of ecolabelling. *Organization, 13*(3), 345–367.

Bowen, F., Cousins, P., Lamming, R., & Faruk, A. (2001). The role of supply management capabilities in green supply. *Production and Operations Management, 10*(2), 174–189.

Boyd, D. E., Spekman, R. E., Kamauff, J. W., & Werhane, P. (2007). Corporate Social Responsibility in global supply chains: A procedural justice perspective. *Long Range Planning, 40,* 341–356.

Bremer, J., & Udovich, J. (2001). Alternative approaches to supply chain compliance monitoring. *Journal of Fashion Marketing and Management: An International Journal, 5*(4), 333–352.

Cashore, B., Auld, G., Bernstein, S., & McDermott, C. (2007). Can non-state governance "ratchet up" global environmental standards? Lessons from the forest sector. *Review of European Community & International Environmental Law, 16*(2), 158–172.

Cerruti, C., Mena, C., Skipworth, H., & Tavoletti, E. (2016). Characterizing agile supply partnerships in the fashion industry. *International Journal of Operations & Production Management, 36*(8), 923–947.

Ciliberti, F., de Haan, J., de Groot, G., & Pontrandolfo, P. (2009). Codes to coordinate supply chains: SMEs' experiences with SA8000. *Supply Chain Management: An International Journal, 14*(2), 117–127.

Ciliberti, F., de Haan, J., de Groot, J., & Pontrandolfo, P. (2011). CSR codes and the principal-agent problem in supply chains: Four case studies. *Journal of Cleaner Production, 19,* 885–894.

Clean Clothes Campaign. (2016). *Dangerous delays on workers safety*. https://cleanclothes.org/resources/publications/dangerous-delays-on-workersafety/view.

Clemens, B., Bamford, C. E., & Douglas, T. J. (2008). Choosing strategic responses to address emerging environmental regulations: Size, perceived influence and uncertainty. *Business Strategy and the Environment, 17,* 493–511.

Collins, D. (2004). Five levees for improving ethical performance. *Strategic Finance, 88,* 19–21.

Das, T. K., & Teng, B. (1998). Between trust and control: Developing confidence in partner cooperation in alliances. *Academy of Management Review, 23,* 491–512.

Delbufalo, E. (2012). Outcomes of inter-organizational trust in supply chain relationships: A systematic literature review and a meta-analysis of the empirical evidence. *Supply Chain Management: An International Journal, 17*(4), 377–402.

Delbufalo, E., & Bastl, M. (2018). Multi-principal collaboration and supplier's compliance with codes-of-conduct. *International Journal of Logistics Management* (forthcoming).

Diller, J. (1999). A social conscience in the global marketplace? Labor dimensions of codes of conduct, social labeling and investor initiatives. *International Labor Organizations, 138,* 99–129.

DiMaggio, P. J., & Powell, W. W. (1983). The iron cage revisited: Institutional isomorphism and collective rationality in organizational fields. *American Sociological Review, 48,* 147–160.

Dyer, J. H. (1996). Does governance matter? Keiretsu alliance and asset specificity as sources of Japanese competitive advantage. *Organization Science, 7,* 649–666.

Egels-Zandén, N. (2007). Suppliers compliance with MNCs codes of conduct: Behind the scenes at Chinese toy suppliers. *Journal of Business Ethics, 75,* 45–62.

Egels-Zandén, N. (2014). Revisiting supplier compliance with MNC codes of conduct: Recoupling policy and practice at Chinese toy suppliers. *Journal of Business Ethics, 119,* 59–75.

Eisenhardt, K. M. (1989). Agency theory: An assessment and review. *Academy of Management Review, 14*(1), 57–74.

Elkington, J. (1998). *Cannibals with forks: The triple bottom line of the 21st Century*. Stoney Creek, CT: New Society Publishers.

Emmelhainz, M., & Adams, R. (1999). The apparel industry response to 'Sweatshop' concerns—A review and analysis of codes of conduct. *Journal of Supply Chain Management, 35*(3), 51–57.

Erwin, P. M. (2011). Corporate codes of conduct: The effects of code content and quality on ethical performance. *Journal of Business Ethics, 99,* 535–548.

Esbenshade, J. (2004). *Monitoring sweatshops: Workers, consumers, and the global apparel industry*. Philadelphia: Temple University Press.

Fairclough, G. (2007). China-product scare hits home, too. *The Wall Street Journal*. http://online.wsj.com/article/SB118487371421671957.html?mod=dist_smartbrief.

Fernandez, A. S., & Chiambaretto, P. (2016). Managing tensions related to information in coopetition. *Industrial Marketing Management, 53,* 66–76.

Fransen, L. W., & Kolk, A. (2007). Global rule-setting for business: A critical analysis of multi-stakeholder standards. *Organization, 14*(5), 667–684.

Frenkel, S., & Scott, D. (2002). Compliance, collaboration, and codes of labor practice: The Adidas connection. *California Management Review, 45*(1), 29–49.

Frundt, H. J. (2000). Models of cross-border organizing in Maquila industries. *Critical Sociology, 26*(1/2), 36–55.

Fynes, B., & Voss, C. (2002). The moderating effect of buyer-supplier relationships on quality practices and performance. *International Journal of Operations and Production Management, 22*(6), 589–613.

Gimenez, C., & Tachizawa, E. M. (2012). Extending sustainability to suppliers: A systematic literature review. *Supply Chain Management, 17*(5), 531–543.

Gundlach, G. T., Achrol, R. S., & Mentzer, J. T. (1995). The structure of commitment in exchange. *Journal of Marketing, 59*(1), 78–92.

Hallett, T. (2010). The myth incarnate: Recoupling processes, turmoil, and inhabited institutions in an urban elementary school. *American Journal of Sociology, 75,* 52–74.

Hardy, K. G., & Magrath, A. J. (1989). Dealing with cheating in distribution. *European Journal of Marketing, 23*(2), 123–129.

Haufler, V. (2003). New forms of governance: Certification regimes as social regulations of the global market. In C. Elliott, E. Meidinger, & G. Oesten (Eds.), *Social and political dimensions of forest certification* (pp. 237–247). Remagen-Oberwinter: Forstbuch.

Hill, R. P., & Rapp, J. M. (2014). Codes of ethical conduct: A bottom-up approach. *Journal of Business Ethics, 123,* 621–630.

Jiang, B. (2009a). Implementing supplier codes of conduct in global supply chains: Process explanations from theoretic and empirical perspectives. *Journal of Business Ethics, 85,* 77–92.

Jiang, B. (2009b). The effects of interorganizational governance on supplier's compliance with SCC: An empirical examination of compliant and non-compliant suppliers. *Journal of Operations Management, 27,* 267–280.

Kelly, E., & Dobbin, F. (1998). How affirmative action became diversity management. *American Behavioral Scientist, 41*(7), 960–984.

Koçer, R. G., & Fransen, L. (2009). Codes of conduct and the promise of a change of climate in worker organization. *European Journal of Industrial Relations, 15*(3), 237–256.

Kondej, M. (2016). *Decoding sustainability in the denim industry: Interview with Michael Kobori, Vice President of Sustainability at Levi Strauss & Co.* Apparel and Footwear in the United Kingdom, Report Euromonitor International.

Kostova, T., & Zaheer, S. (1999). Organizational legitimacy under conditions of complexity: The case of the multinational enterprise. *Academy of Management Review, 24*(1), 64–81.

Krause, D. R., Handfield, R. B., & Tyler, B. B. (2007). The relationships between supplier development, commitment, social capital accumulation and performance improvement. *Journal of Operations Management, 25*(2), 528–545.

Lenox, M. J., & Nash, J. (2003). Industry self-regulation and adverse selection: A comparison across four trade association programs. *Business Strategy and the Environment, 12,* 343–356.

Lim, S. J., & Phillips, J. (2008). Embedding CSR values: The global footwear industry's evolving governance structure. *Journal of Business Ethics, 81,* 143–156.

Locke, R., Kochan, T., Romis, M., & Qin, F. (2007). Beyond corporate codes of conduct: Work organization and labour standards at Nike's suppliers. *International Labour Review, 146*(1–2), 21–40.

Lund-Thomsen, P. (2008). The global sourcing and codes of conduct debate: Five myths and five recommendations. *Development and Change, 39*(6), 1005–1018.

Lund-Thomsen, P., & Nadvi, K. (2011). Cluster, chains and compliance: Corporate social responsibility and governance in football manufacturing in South Asia. *Journal of Business Ethics, 93,* 201–222.

McEvily, B., & Marcus, A. (2005). Embedded ties and the acquisition of competitive capabilities. *Strategic Management Journal, 26,* 1033–1055.

Mele, V., & Schepers, D. H. (2013). E pluribus unum? Legitimacy issues and multi-stakeholder codes of conduct. *Journal of Business Ethics, 118,* 561–576.

Milgrom, P., & Roberts, J. (1992). *Economics, organization and management*. New Jersey: Prentice Hall Inc.

Moore, G. (2001). Corporate social and financial performance: An investigation in the U.K. supermarket industry. *Journal of Business Ethics, 34*(3/4), 299–315.

Moran, P. (2005). Structural versus relational embeddedness: Social capital and managerial performance. *Strategic Management Journal, 26,* 1129–1151.

Murry, J. P., & Heide, J. B. (1998). Managing promotion program participation within manufacturer retailer relationships. *Journal of Marketing, 62*(1), 58–68.

Nelson, R. R., & Winter, S. G. (1982). *An evolutionary theory of economic change*. Cambridge: Belknap Press/Harvard University Press.

O'Rourke, D. (2002). Monitoring the monitors: A critique of corporate third party labor monitoring. In R. Jenkins, R. Pearson, & G. Seyfang (Eds.), *Corporate responsibility and ethical trade: Codes-of-conduct in the global economy* (pp. 196–208). London: Earthscan.

Pedersen, E. R., & Andersen, M. (2006). Safeguarding Corporate Social Responsibility (CSR) in global supply chains: How codes of conduct are managed in buyer-supplier relationships. *Journal of Public Affairs, 6*(3/4), 228–240.

Plambeck, E., Lee, H. L., & Yatsko, P. (2012). Improving environmental performance in your chinese supply chain. *MIT Sloan Management Review, 53*(12), 43–51.

Porter, T., & Ronit, K. (2006). Self-regulation as policy process: The multiple and criss-crossing stages of private rule-making. *Policy Sciences, 39*(1), 41–72.

Prakash, A., & Potoski, M. (2007). Collective action through voluntary environmental programs: A club theory perspective. *Policy Studies Journal, 35*(4), 773–792.

Radaelli, C. (2003). The code of conduct against harmful tax competition: Open method of coordination in disguise? *Public Administration, 81*(3), 513–531.

Radin, T. J. (2004). The effectiveness of global codes of conduct: Role models that make sense. *Business and Society Review, 109*(4), 415–447.

Risso, M. (2012). A horizontal approach to implementing corporate social responsibility in international supply chains. *International Journal of Technology Management, 58*(1/3), 64–82.

Roberts, D., & Engardio, P. (2006, November 27). Secrets, lies, and sweatshops. *Business-Week*.

Roloff, J. (2007). Learning from multi-stakeholder networks: Issue focused stakeholder management. *Journal of Business Ethics, 82*(1), 233–250.

Ross, R. J. S., & Chan, A. (2002). From North–South to South–South: The true face of global competition. *Foreign Affairs, 81*(5), 8–13.

Schäferhoff, M., Campe, S., & Kaan, C. (2009). Transnational public-private partnerships in international relations: Making sense of concepts, research frameworks, and results. *International Studies Review, 11*(3), 451–474.

Scott, W. R. (2008). *Institutions and organizations. Ideas and interests* (3rd ed.). Thousand Oaks, CA: SAGE.

Sethi, S. P. (2002). Standards for corporate conduct in the international arena: Challenges and opportunities for multinational corporations. *Business and Society Review, 107*(1), 20–40.

Short, J. L., Toffel, M. W. & Hugill, A. L. (2015). Monitoring global supply chains. *Working Paper, Harvard Business Review* 0–32.

Stafford, S. (2007). Should you turn yourself in? The consequences of environmental self-policing. *Journal of Policy Analysis and Management, 26*(2), 305–326.

Taylor, M. (2011). Race you to the bottom … and back again? The uneven development of labour codes of conduct. *New Political Economy, 16*(4), 445–462.

Tidström, A. (2014). Managing tensions in coopetition. *Industrial Marketing Management, 43*(2), 261–271.
Tomkins, C. (2001). Interdependencies, trust and information in relationships, alliances, and networks. *Accounting, Organizations and Society, 26,* 161–191.
Williamson, O. E. (1985). *The economic institutions of capitalism: Firms, markets and relational contracting*. New York, NY: Free Press.
World Commission on Environment and Development. (1987). *Our common future*. New York, NY: Oxford University Press.

Chapter 3
Agency Theory and Sustainability in Global Supply Chain

Abstract The third chapter discusses the contribution of corporate responsibility collaborations among competitors. The study explains the origin of agency problems in global supply chain and the negative impact of opportunistic behaviours on sustainable development. After that, it addresses the effects of collaboration among competitors on the degree of suppliers' compliance and analyses how this collaboration shapes the code of conduct's monitoring system and relationship management.

Keywords Agency theory · Collaboration · Compliance · Monitoring · Safeguards

3.1 An Agency Theory Perspective on Buyer–Supplier Relationships in Global Supply Chain

While companies from Western countries have increased the use of outsourcing and sourcing from low-cost economies, a wide range of stakeholders has shown concerns about the social and environmental impacts of production and consumption. In such contexts, the interest in issues related to social responsibility, sustainable sourcing, environmental management and green supply chains constantly increases (Aguinis and Glavas 2012; Hill and Rapp 2014; Lee and Kim 2009). As firms increasingly rely on extended networks of suppliers to serve their customers, the concept of CSR is also being extended from an individual firm's domain to a set of actions developed along the entire supply chain (Amaeshi et al. 2008; Andersen and Skjøtt-Larsen 2009).

Considering sustainability issues in supply chain contexts implies that the responsibility for the inter-firm division of labour, partners' performance, as well as control and monitoring of supply chain activities should be taken over by key actors, generally MNCs (Gereffi 2001). Given the MNCs' power and coordinating role within global supply chains, these corporations are considered to be responsible for the environmental and social practices of their partners (Jenkins 2001; Maloni and Brown 2006). When the success of CSR initiatives depends on the actions and behaviour

E. Delbufalo, *Agency Theory and Sustainability in the Global Supply Chain*, SpringerBriefs in Business, https://doi.org/10.1007/978-3-319-72793-6_3

of multiple actors in the supply chain, the governance of these inter-organisational relationships is of critical importance (Sobczak 2006).

In order to extend sustainability practices to the most critical partners in a global supply chain (i.e. the suppliers, mostly when they operate in emerging economies), a growing number of MNCs engage in collaborative efforts with other firms. Sometimes, the promoters of such collaboration are competitors in the same industries, which share similar concerns about the social, ethical and environmental performance of their supply base (Appolloni et al. 2013). One of the main purposes of these collaborative efforts is to develop a shared code of conduct aimed towards managing and monitoring the suppliers' compliance with ethically and socially responsible practices (Arya and Salk 2006; Mele and Schepers 2013). These codes define sustainable operational issues along with some other requirement; for example, working conditions must be safe and hygienic, child labour is not used, working hours are not excessive and the wages and pay conditions are fair (Jiang 2009).

However, anecdotal evidence shows a consistent lack of commitment and non-compliance with social and environmental requirements by suppliers, even if the collaborative initiatives developed by MNCs are progressively growing in number and extension (Ciliberti et al. 2009). In fact, "active commitment is a precondition for the successful implementation of the [collaborative initiatives], but the incentive to comply with [them] does not necessarily extend to all the actors in the chain" (Pedersen and Andersen 2006: p. 228). Despite the proliferation of shared codes of conduct and monitoring activities developed collaboratively by a multiplicity of actors—not only competitors but also NGOs, governmental agencies, labour union representatives, etc.—the degree of suppliers' non-compliance is still high. Moreover, recent research has reported difficulties in managing a multiplicity of conflicting interests among MNCs and non-compliant suppliers, also questioning the legitimacy of such efforts and their potential benefits (Mele and Schepers 2013). These negative aspects can lead MNCs into a high-risk situation that can result in consumer sanctions, negative press, capital loss, government interventions and damage to the company's brand (Aguinis and Glavas 2012).

The intent of this chapter is to present the main characteristics of collaborative initiatives for sustainability developed by MNCs in global supply chains. By adopting an agency theory perspective (Eisenhardt 1989; Mitnick 1973), the chapter shows the effects of a collaboration among MNCs on suppliers' degree of compliance with (shared) codes of conduct. It also shows how MNCs can effectively manage the relationships with non-compliant suppliers in order to safeguard themselves against negative returns. Agency theory provides a novel perspective on global supply chain sustainability problems as it is specifically focused on the relationship between principal and agent—which are engaged in cooperative behaviour but have differing goals and asymmetric information. This situation is mirroring the one developed by MNCs (acting as principals) and suppliers (acting as agents) where the degree of opportunism and informative asymmetry translates into non-compliant behaviours. The study enriches the theoretical arguments with some notable experiences of industry-specific collaboration among competitors for sustainable supply chain management and code of conduct implementation.

3.1.1 Sustainability and Agency Problems in Global Sourcing

Agency theory is concerned with the study of problems that arise when one party (the principal) delegates authority—in terms of decision-making and control over specific tasks or activities—to another party (the agent) (Eisenhardt 1989; Mitnick 1973).[1] The theory focuses on relationships in which principal and agent are engaged in cooperative behaviour, but have differing goals and attitudes towards risk. Because the unit of analysis is the contract governing the relationship between the principal and the agent, the focus of agency theory is on determining the most efficient contract governing the principal–agent relationship, given the assumptions about people (self-interest, bounded rationality, risk aversion), organisations (goal conflict or incongruence) and information (information asymmetry between principal and agent) (Eisenhardt 1989). An overview of agency theory is given in Table 3.1.

Agency problems are essential problems of the principal, and they arise because the agent has private information about the principal's intentions, abilities and behaviour. Two agency problems exist: misrepresentation of ability (namely, adverse selection) and lack of effort (namely moral hazard), both of which are attributed to the agent (Saam 2007). Given as fixed the assumptions about people, when an agency problem exists, principals have two basic choices (Whipple and Roh 2010):

- to reduce/eliminate goal conflict, which is responsible for the motivation to cheat;
- to reduce/eliminate information asymmetry, which is responsible for the ability to cheat.

In order to address the goal conflict issue (or motivation to cheat), the principal must incentivise the agent in a way that encourages the latter to act consistently with the former's interests. In other words, the incentive should act in a manner to align the goal of both the principal and the agent (Eisenhardt 1989; Jensen and Meckling 1976). Incentives could be offered in the form of an outcome-based contract whereby "the agent is compensated solely on the basis of performance outcomes" (Lassar

Table 3.1 Overview of agency theory

Key idea	Principal—agent relationships should reflect efficient organisation of information and risk-bearing costs
Unit of analysis	Contract between principal and agent
Human assumptions	Self-interest; bounded rationality; risk aversion
Organisational assumptions	Goal conflict and information asymmetry between principal and agent
Contracting problems	Agency problems: moral hazard and adverse selection
Problem domain	Relationships in which the principal and the agent have partly differing goals and risk preferences

Source Adapted from Eisenhardt (1989)

[1]See Chap. 1 for additional details.

and Kerr 1996: p. 615), or behaviour-based contracts, which "identify the specific behaviours expected of the agent" (Lassar and Kerr 1996: p. 616). In the context of buyer–supplier relationships, outcome-based contracts would typically involve some type of price premium in order to "make cheating unprofitable" (Mishra et al. 1998: p. 280), such as cost-plus incentives and/or risk sharing mechanisms. Behaviour-based contracts consider—and detail—specific behaviours required by the buyer and "may impose restrictions on variables such as inventory levels, range of products stocked [...]" (Lassar and Kerr 1996: p. 616).

In order to reduce/eliminate the information asymmetry (or ability to cheat), the principal can utilise a behaviour-based contract in an effort to stimulate the agent's appropriate behaviour (Eisenhardt 1989). The underlying assumption is that the expected behaviour will lead to the expected outcomes, and as such "if the principal can ensure that the agent has met behavioural expectations, the principal can assume that the expected performance outcomes will result" (Whipple and Roh 2010: p. 344).

Both behaviour-based and outcome-based contracts are associated with (agency) costs, strictly commensurate with a number of relationship-specific factors, such as the complexity of the task, the ability to monitor and the degree of uncertainty (Arrow 1985). These costs arise because both parties in the relationship (the principal and the agent) are assumed to hold self-serving interests and the contracts used to align those interests are per se incomplete. Moreover, both contract types imply monitoring activities—which are costly and time consuming—to verify respectively the agent's behaviour or the performance outcomes.

The incompleteness and the agency costs linked to contractual mechanisms challenge the possibility for the principal to substantially reduce information asymmetry and goal conflict by characterising an optimal contractual form to manage the relationship with the agent. As a consequence, the principal requires additional mechanisms to mitigate the enduring agency problems. Agency theory literature suggests two specific actions: improve the information acquisition process and develop effective relationship management (Eisenhardt 1989; Jensen and Meckling 1976).

- *Information acquisition process*: This process serves to monitor the agent's behaviour and performance. It is recognised as the most powerful tool for the reduction of information asymmetry (Jensen and Meckling 1976). The information acquisition process—and the corresponding monitoring choices (i.e. internal vs. external monitoring)—allows the principal to systematically collect data and information on the agent's activities. The positive return of such activity can be synthesised as follows: the agent is more likely to behave in the interests of the principal when the principal has information to verify the agent's behaviour (Eisenhardt 1989). However, the development of an effective information acquisition process generates costs for the principal and it is subject to some degree of uncertainty as perfect monitoring is unrealistic. As pure information monitoring systems are often not efficient, they may be complemented by safeguards and protective mechanisms.

- *Relationship management and safeguards*: Agency theory provides suggestions about the contractual arrangement of a buyer–supplier relationship along with the development of safeguards to reduce the threat of opportunistic behaviours. In particular, safeguards are defined as the "means to ensure that an agent fulfils the obligations according to the agreement" (Koch 1995: p. 8). In other words, safeguards can be considered as a mechanism to increase the efficacy of an economic exchange by making it less convenient for the agent to deceive the principal's expectations. The common rationale for safeguarding a relationship is that "the agent functions as a utility maximiser and the safeguards therefore all contain some aspect of formal, standardised and foreseeable economic incentive, either positive or negative" (Eriksen and Jessen 2009: p. 32). The safeguards are a means to reduce goal conflict by making it economically prudent not to behave opportunistically and by making the principal able to impose serious economic consequences on the agent in case the defection happens (Buvik and Reve 2002). Typical examples of relationship management through safeguards are: rewards and sanctions, and reputation effect (Bergen et al. 1992).

3.1.2 Opportunism and Implementation of Code of Conduct

Opportunism poses a threat to inter-firm relationships, such as those regarding the planning and implementation of a code of conduct in global supply chains (Das and Rahman 2002; O'Donnell 2000). The opportunistic assumption (e.g. Williamson 1993) becomes relevant in this specific context as these activities "can be costly and time consuming" (Pedersen and Andersen 2006: p. 230; Kapstein 2001; Kolk 2000; Utting 2000). Consequently, the supplier might have an economic incentive to reduce social and environmental standards in order to achieve financial gains (Pedersen and Andersen 2006). Moreover, the distribution of costs and perceived benefits associated with code of conduct compliance can exacerbate the agency problems. In fact, the perceived benefits associated with social responsibility practices are higher for the firm that is closer to its customers—that is also the initiator of the code—than for the other actors in the supply chain. The initiator (or the buying firm), having a wider known brand, can more easily communicate with consumers and is thus more affected by their attitude and judgment than suppliers further upstream of the chain. The suppliers will just have to share the indirect benefits, for example, from additional deliveries and employee satisfaction. While the positive publicity and return associated with a code of conduct initiative will benefit the initiator more than the suppliers, the costs associated with complying with the code are equally distributed among all the supply chain actors (Barrientos 2002). This creates a situation where the buyer–initiator obtain most of the benefits while the supplier is left with the costs. "Such an imbalance, if not dealt with by the buyer, is a strong cause of opportunistic behaviour" (Eriksen and Jessen 2009: p. 29).

Agency theory suggests that the threat of opportunism can be reduced by formulating an enforceable contract, which includes all contingencies (Greve 2000;

Milgrom and Roberts 1992). Codes of conduct—as with many other contracts—are incomplete and do not accomplish this goal. Coherently with this position, Sethi (2002) criticises codes of conduct for their vague formulation and for being typically presented as public statements of intent and purpose without any specific content. Codes of conduct have also been characterised by a remarkable lack of efficient monitoring systems, which further reduces the possibility of controlling their compliance by the suppliers (Emmelhainz and Adams 1999; Lee and Kim 2009; Pedersen and Andersen 2006). Furthermore, suppliers entering into contractual relationships with large companies may be confronted with multiple codes that incorporate the different standards and requirements they need to comply with (Andersen and Skjøtt-Larsen 2009). This entails additional costs for suppliers and thus the additional threat of opportunism.

To deal with the agency problems and to increase the suppliers' compliance with the code of conduct, some large companies are promoting collaborative initiatives among competitors to facilitate the identification of common standards across their global supply chain (Appolloni et al. 2013; Risso 2012). These collaborative efforts lead to the development of collective CSR strategies and to the formulation of shared practices, resulting from the combination and integration of the individual (i.e. firm level) code of conduct. The positive effect of collaboration among competitors on CSR strategy implementation is an outstanding management issue although—with few exceptions (Arya and Salk 2006; Delbufalo and Bastl 2018; Risso 2012)—still largely unexplored.

3.2 Corporate Responsibility Collaboration and Supplier's Compliance

A corporate responsibility coalition can be defined as the "joint mobilization of resources and formulation of actions set up by organisations sharing common perspectives and needs about environmental and social issues" (Risso 2012: p. 66). Some scholars refer to this situation as "horizontal inter-chain collaboration" to distinguish the phenomenon from the traditional "vertical collaboration" which is likely to appear between each large company and its suppliers (e.g. Risso 2012: p. 68; Peloza and Falkenberg 2009).

In many cases, companies are collaborating with competitors as well as NGOs and public sector bodies, to address specific problems. For example, in the "Refrigerants, Naturally!" coalition, Coca-Cola, PepsiCo, Red Bull and Unilever established an alliance with Greenpeace and UNEP to develop more sustainable refrigeration technologies (Grayson 2014). A corporate responsibility coalition—that can be recognised as a collaboration among competitors for sustainability purposes—generally gives rise to the establishment of common standards and a common set of CSR policies and action plans. The shared policies, jointly developed by a multiplicity of large buyers in a specific industry, are likely to encourage the development of best practices and would avoid exposing suppliers to potentially conflicting requirements. In such cases, "the bilateral relationships between suppliers and large enterprises would be informed by uniform standards creating a level playing field among all actors active in a specific sector" (Risso 2012: p. 68).

As suggested by Grayson (2014): "General corporate responsibility coalitions have been supplemented by coalitions focused on particular sectors, or on specific issues including water or human rights". Some of the leading international corporate responsibility coalitions, such as Business for Social Responsibility (BSR), CSR Europe and the World Business Council for Sustainable Development, are involved in specialist coalitions, such as Better Coal (BSR), looking at CSR in the coal supply chain; European Automotive Working Group on Supply Chain Sustainability (CSR Europe); and The Tire Industry Project, which looks at the environmental impact of tyre making and use.

The benefits of a collaboration among competitors for sustainability also arise from another concern: large corporations faced with a multiplicity of suppliers around the world need to sustain high costs and have time to monitor (i.e. audit) their partners' social and environmental performance. Considering that large companies (buyers) in a specific industry might share a large portion of their suppliers, a cross-industry collaboration among them might result in less costly and more effective CSR implementation practices. This has been so in several cases, such as the Business Social Compliance Initiative (BSCI) and the Global Social Compliance Program (GSCP) in the consumer goods industry, as well as the Responsible Jewellery Council (RJC) in the jewellery industry (for details see Risso 2012). As reported in the September 2017 report by the Network for Business Sustainability, for many MNCs a coalition among competitors is nothing new: for example, Nike is a founding member of both the Sustainable Apparel Coalition and the Zero Discharge Hazardous Chemicals (ZDHC). In response to the Greenpeace Detox campaign, ZDHC has brought together competing brands such as H&M, Nike and Zara to work on eliminating the use of the 11 most hazardous chemicals in textile production by 2020. ZDHC members compiled a material restricted substance list, initiated supplier training, expanded membership to include suppliers and other stakeholders and established ZDHC as a separate entity in 2015. BASF and Dow Chemicals, both long-standing members of the Responsible Care programme, also know the benefits of working with their competitors. Responsible care has been bringing competitors together to solve issues around factory compliance, safety and product stewardship—and has been doing so for 30 years and counting (NBS 2017). Boxes 1 and 2 provide details about two interesting examples of corporate coalition for sustainability: the Joint Audit Cooperation (JAC) in the telecommunication industry and the Sustainable Apparel Coalition in the apparel sector.

Using an agency theory perspective, one may argue that the establishment of a corporate responsibility coalition (and the consequent development of shared standards and practices) leads to an increased economic and contractual control over the suppliers and the entire supply chain, resulting in more opportunities for compliance. The motivation behind the increase in economic and contractual control over the supply base is strictly linked to the concept of the bargaining power of the buyers over the suppliers. More arguments on this point follow.

Bargaining power is a combination of "market power" and "resource power" (Buvik and Reve 2002). Market power relates to an actor's ability to influence another actor due to its economic importance or the degree of market concentration. Resource power is not dependent on size or direct economic dominance, but is rather determined as a result of the importance that the relationship has for the parties involved. A high degree of resource power appears if a buyer can be of special importance to a supplier due to access to valuable know-how, distinctive capabilities or market information. If a supplier can be made dependent on the continuity of a relationship with a buyer, the buyer will have a stronger position when trying to influence the supplier to, for example, follow CSR requirements (Prosman et al. 2016). The original formulation of agency theory addresses the power issue by designing the right contract in such a way that additional considerations of power asymmetries should become redundant (Perrow 1986). Agency theory's assumption of a risk-averse agent (Eisenhardt 1989) locks the role of bargaining power into a static explanation in which the principal—due to its more neutral attitude towards risk—is less dependent on the agent, while the agent—due to its risk aversion—is more dependent on the principal (Tomkins 2001). However, in a buyer–supplier relationship this issue may be different (Eriksen and Jessen 2009). First, unlike the traditional agency theory logic, suppliers are not necessarily risk averse: suppliers could diversify their risks through their relationships with a range of different buyers that they supply within the same industry. The supplier, not highly affected by a single buyer, has less incentive to follow the shared standards because the risk of not doing so is virtually diversified away. Second, it may not be simple to transfer risk from the buyer to the supplier in a way that will leave the buyer exempt, as assumed in traditional agency theory. "This problem builds on the rationale that any code of conduct violation by the supplier will affect the buyer at least as much as it can ever affect the supplier" (Eriksen and Jessen 2009: p. 30). This is due to the assumption that the exposure of non-compliance issues causes most damage to the brand company closer to the end-consumer than the supplier further upstream in the supply chain (Barrientos 2002).

Following this reasoning, Delbufalo and Bastl (2018) argue that an important driver for non-compliance with a code of conduct is the limited bargaining power of the initiator and the consequent limited ability to comply with the risk attitude of the suppliers. In this context, a corporate responsibility collaboration—and shared standards—may have a strong impact on buyers' bargaining power. As the initiator of a code of conduct is often a multiplicity of large buyers (competitors in the same industry), the possibility of diversifying the risk of non-compliance by the suppliers decreases. Equally, the negative response to CSR requirements by each supplier may result in collective and more coordinated corrective actions promoted by the coalition of buyers and in damage to business relationships with multiple large buyers. By developing shared standards and practices, the multiple buyers present themselves to the suppliers as a unique organisational entity with regard to social and environmental issues, having a larger market power and an increased ability to influence the suppliers due to their collective economic importance. In other words, a multi-principal collaboration determines a kind of "market concentration" for the suppliers, thus increasing the bargaining power of the buyers (Delbufalo and Bastl 2018). Through increased bargaining power, the promoters of the collective CSR strategy exert a stronger influence over the suppliers to follow sustainability requirements and thus reduce the threat of non-compliance.

The following sections address the impacts of a corporate responsibility coalition on two specific activities—information acquisition process and relationship management—developed by MNCs in order to reduce the negative return of agency problems in global supply chains.

Box 1—Corporate responsibility coalition in the telecommunication industry

A notable example of corporate coalition among competitors is the Joint Audit Cooperation (JAC), a consortium of MNCs operating in the telecommunication industry (http://jac-initiative.com). This association aims to verify, assess and develop CSR implementation across the manufacturing centres of the most important ICT multinational suppliers. JAC members share resources and best practices to develop long-term CSR implementation in the different layers of the ICT supply chain at an international level. The main objectives of the JAC project are to: (1) increase efficiency and productivity for companies and suppliers; (2) contribute to globally responsible development and ensure continuity of supply by understanding and reducing CSR risks and (3) enhance corporate image and reputation for all JAC members (Appolloni et al. 2013). The origin of JAC dates back to 2010 when Telecom Italia, Deutsche Telekom and France Telecom signed a Memorandum of Understanding (MoU) called the "CSR Joint Audit Cooperation". Belgacom, KPN, Swisscom and Vodafone subsequently joined the project in September 2011. Between 2012 and 2016 further members joined becoming part of the initiative (Verizon, Orange, Telenor and TeliaSonera). As a result of this agreement the telecom operators will develop shared standards and practices to carry out joint audits of environmental and social sustainability among common suppliers in production plants located in Asia (i.e., China, Taiwan, India, Japan, South Korea). The companies share a relevant part of their suppliers, mostly those located in China (Shenzhen city of the Guangdong province) (Appolloni et al. 2013).

The JAC members account for a large part of the delivery volumes of the Asian suppliers. In fact, a recent estimation suggests that more than 20% of the sourcing activities/volumes made by the telecommunication companies in the Asian countries are undertaken by JAC members. Besides, for many small and medium suppliers the JAC members are the most important (in some case the exclusive) buyers of their products. This concentration of the market in Asian suppliers reduces the possibility for them to diversify the risks through their relationships with a range of different buyers that they supply. The JAC initiative has significantly impacted on both the market power and the resource power of the buyers. On the one hand, the collaboration among telecom companies has increased the market power, by increasing the suppliers' dependence on the continuity of multiple relationships with all the JAC members. In doing so, it has enhanced the buyers' ability to influence the suppliers due to their economic importance, thus increasing the probability of suppliers' compliance with the shared CSR standards. On the other hand, the collaboration has improved the capabilities of JAC members to build up standard methodologies for auditing and monitoring, improve communication with the suppliers and develop innovative procedures and sourcing practices. These benefits have increased the resource power of JAC members, as they appear to be of particular importance to the Asian suppliers by providing access to their valuable and jointly developed know-how, market information and effective supply chain management. The increase in buyers' bargaining power is particularly evident if one considers that a certain degree of competition is growing among suppliers (especially the small ones) to quickly comply with the CSR standards and become socially responsible and reliable partners for all JAC members.

Box 2—The Sustainable Apparel Coalition

The Sustainable Apparel Coalition (SAC) is a collaboration among competitors in the apparel, footwear and textile industry for sustainable production in global supply chains. The aim of the coalition—which accounts for more than 200 global members with a combined annual revenue exceeding $500 billion—is to eliminate production processes or activities that produce unnecessary environmental harm and potentiate the activities having a positive impact on the people and society (https://apparelcoalition.org/the-sac). The vision of the coalition can be synthesised by using the members' words: "Collaboration is the heartbeat of the SAC. No company alone can shift the existing industry paradigm. To ignite the change required to redefine how the industry is run, peers and competitors come together as a united front, adhering to the Coalition's set of core collaboration values that are designed to further impactful change across the industry. Through SAC membership, brands, retailers, and manufacturers commit to transparency, the sharing of best practices, and making meaningful improvements, a full-circle collaboration that benefits all involved".[a] Although the coalition has been developed from the beginning as a corporate collaboration intra- and inter-industry, progressively a number of different stakeholders have joined the project and given their specialised contribution to the development of the initiative. In fact, membership has been allowed for governmental agencies (The Sustainable Trade Initiative, IDH), NGOs (e.g. Natural Resources Defense Council, NRDC) and service providers and management consulting firms (e.g. CPI2, McKinsey & Company, Outdoor Industry Association, Zero Discharge of Hazardous Chemicals Programme, ZDHC).

The origin of the Sustainable Apparel Coalition dates back in 2009 when Walmart and Patagonia wrote a joint letter inviting CEOs of leading global companies to come together to develop an index that would measure the environmental impact of their products. In 2010, competitors from the apparel industry came together to begin collaborative work on the Higg Index and the coalition was formally established. The Higg Index represents the main outcome of such collaboration along with a number of other tools useful for any company committed to sustainable production and sourcing in the apparel sector. Considering that more than 10,000 manufacturers currently include the Higg Index in their sustainability strategy, the utility to take part to the coalition clearly increases. As in the case of the JAC experience, the concentration of the market realised by the members (buyers and manufacturers in the same industry) reduces the possibility for their respective suppliers to diversify the risks through supply relationships with a range of alternative buyers. In doing so, the incentives for compliance with the SAC's members' requirements increase as it become less convenient—and more risky—for the suppliers to default. At the same time, the tool developed by the collaborators (Higg Index) increases the homogeneity, comparability and transparency of the production processes, thus reducing the informative asymmetry in buyer–supplier relationships.

[a]From the website of the Sustainable Apparel Coalition (https://apparelcoalition.org/collaboration-impact/).

3.3 Information Acquisition Process

Asymmetric information is a central assumption in agency theory. Perfect information reduces the threat of opportunism as it makes it easier to detect actors' opportunistic behaviour (Petersen 1993). In such an ideal situation, the principal would therefore employ a behaviour-based contract and compensate the agent accordingly (Eisenhardt 1989). However, in a buyer–supplier relationship the actors have access to different and often insufficient information about their partners' behaviour. In an international supply chain, information issues are further complicated by cultural and physical distances between actors. If actors in a supply chain relationship are within a close distance, other things being equal, it would be easier (or less costly) to obtain wanted information. Similarly, the greater the distance the harder (or more expensive) it becomes to obtain wanted information (Eriksen and Jessen 2009; Pedersen and Andersen 2006).

Information does not in itself possess the characteristics of a safeguard; it is rather a means necessary to utilise safeguards, to reduce the information asymmetry between parties and thereby enabling the safeguards to function. Though not a safeguard, the information acquisition system is a vital part for the effectiveness of CSR strategies' implementation and for the suppliers' monitoring process (i.e. auditing process). Information can be gathered either by the buyer itself or by a third party. The third party often has unique competences related to information acquisition and elaboration which the buyer does not necessarily possess. This means that the third party is better suited to collect the right information and may do it in a more efficient manner. Furthermore, not having any stake in the relationship, the third party should be impartial when collecting and reporting information. For this reason, stakeholders might find it more reliable to have a third party perform the information gathering (World Bank 2003). However, this solution also has its drawbacks. It is costly and may utilise a standardised procedure without adaptation for a specific context. It also introduces an additional intermediary which could result in a loss of information.

A corporate responsibility collaboration has an impact both on the efficiency and effectiveness of the information acquisition process. To obtain wanted information concerning suppliers' behaviour in different geographic areas is costly and time consuming. This is particularly true if the intervention of a specialised third party with a deeper knowledge of the context and with distinctive competences is needed. The information acquisition process to be effective needs to be employed on a large scale and this implies considerable resource expenditure that a single organisation might not be able to afford. Thus, the decision to join a coalition among competitors could reduce the costs, complexity and risks of searching for wanted information. The collaborating buyers might put together tangible and intangible resources to develop a collective information gathering process, offering themselves as a unique voice to suppliers. Furthermore, collaboration among competitors might be helpful in improving the capabilities of buyers to identify specific procedures (i.e. requirements) and parameters to measure the level of CSR in their supply chain, thus increasing the quality of information collected. By doing so, they also improve the transparency and comparability of the results and can benchmark with existing CSR schemes.

In many cases, the collaborators design a monitoring process based on on-site auditing. The most common criteria adopted in order to legitimate the auditors (either in case of internal or external auditing) and ensure the effectiveness of the process are the following (Appolloni et al. 2013).

- *Cooperative attitude with the suppliers*: the buyers (MNCs) share with the suppliers a deep understanding of supply chain risks and vulnerabilities in such a way that the importance of sustainable and responsible development along with the costs (and damages) of non-compliant behaviours becomes evident to all the partners;
- *Credibility*: the audits—either conducted by independent auditing companies or by internal auditors (i.e. auditors trained and managed by the buyers)—need to be based on widely recognised criteria and measures defined by the collaborating members and largely based on recognised standards such as, for example, SA 8000 (Social Accountability) and ISO 14001 standard (Environmental Management);

- *Confidentiality*: MNCs sign with the suppliers' non-disclosure agreements in order to maintain confidentiality of audit results and ensure that the information collected will only be used within the scope of the initiative and never for opportunistic competitive purposes.

As suggested by Delbufalo and Bastl (2018) and Risso (2012), corporate responsibility collaboration helps to provide all the buyers involved with high-quality information about the suppliers' behaviour. Consequently, with more reliable information available at a lower cost (as the unitary cost for the single buyer is reduced) it would easily be possible to detect suppliers' opportunistic behaviour and employ corrective mechanisms and rewards to compensate the suppliers accordingly. Equally, the suppliers would be discouraged from employing opportunistic behaviours because of the more effective information acquisition process which would give greater evidence of their performance. Thus, the improved information acquisition process and the possibility of disseminating the results of suppliers' performance monitoring should reduce the likelihood of non-compliance of the suppliers. From the above, one can conclude that a corporate responsibility collaboration improves the buyers' information acquisition process and thus reduces the likelihood of non-compliance by the suppliers. Box 3 provides a description of the monitoring process developed by JAC's members to collect information from the Asian suppliers. Box 4 describes the measuring tools developed for monitoring purposes by SAC's members.

Box 3—The monitoring and auditing process in JAC

The JAC's members (large telecom buyers/operators) agree to conduct coordinated on-site audits based on a common assessment methodology, which includes the follow-up of any corrective action plans identified. Each member has the responsibility to operate on behalf of the other members and lead an end-to-end supplier assessment. Suppliers are identified by all the JAC members, and then equally assigned in groups to each member who bears the cost and management of their assigned suppliers, with any subsequent results shared amongst the member companies (Appolloni et al. 2013).

The collaboration has also led to a joint list of suppliers undergoing the audit being drawn up, together with the procedure for the checks to be carried out. The procedure is developed in three steps: (1) the audits are carried out by international and professional companies specialised in CSR auditing; (2) suppliers are asked to sign confidentiality agreements on the results of the audits, which are reserved for JAC members only; and (3) following the audits, corrective plans are drawn up with the suppliers, which are then appropriately monitored by JAC members (for details see JAC Guidelines available at: http://jac-initiative.com/how-it-works/#audit).

The improvement of the quality and quantity of information about the Asian suppliers' social performance can be considered one of the most important objectives of the JAC collaboration. JAC members share the belief that an efficient information acquisition system is a vital part for the implementation of CSR strategy. Information is essential to evaluate the social performance of suppliers, their reliability as transaction partners, finally reducing the uncertainty and risks of global sourcing and supply chain management. JAC members have also jointly developed a checklist reporting CSR parameters and measures covering five areas: labour, health and safety, environmental, ethics, management systems. The audit process will involve ten stages: opening meeting, facility evaluation, documentation review, management interviews, worker interviews, closing meeting, audit report, supplier management, response, follow-up verification (JAC 2017).

The independent auditing companies (third party) involved in the auditing process have to perform on-site verification to identify the level of conformity with the CSR principles and to define areas for improvement. JAC members considered these companies to be better suited for collecting the right information in a more efficient manner. The impartiality of the auditing companies and their specialised competencies and skills has led to a significant improvement in the quality of information gathered. Furthermore, the timing and cost for collecting the buyers' behaviour information have been significantly reduced from the beginning. Besides, the reduction of unitary costs for the information acquisition has allowed the JAC members to employ this procedure on a large scale: they are currently auditing more than 140 Asian suppliers but this number is expected to grow fast in the next years.

Box 4—The monitoring tools in SAC

Developed by the Sustainable Apparel Coalition, the Higg Index is a suite of tools that enables brands, retailers and facilities of all sizes—at every stage in their sustainability journey—to accurately measure and score a company or product's sustainability performance (for details see https://apparelcoalition.org/the-sac). The Higg Index delivers a systemic overview of the social and environmental impact of production processes and empowers all the actors involved (not only buyers but also suppliers) to make improvements that protect the work conditions, the environment and local communities. The Index is organised in three different sections which correspond to a set of specific tools: product, facility and brand tools. Higg Index Product Tools help suppliers and manufacturers understand the sustainability impact of apparel, footwear and textiles products by means of a set of metrics measuring the performance of the design process (with specific regard to the materials and components adopted) as well as the environmental impact of production and distribution. The Higg Facility Tools report a set of metrics for monitoring and assessing the social and environmental performance of the facilities. The tools allow a benchmarking by facility type and make it possible for the facility managers to compare their performance against that of their peers. Finally, the Higg Brand and Retail Tools measure the environmental and social impacts of the operations, assessing a product's lifecycle structure from materials sourcing through to its end of use. To build credibility and comparability of Higg Index scores, the SAC develops a verification programme for each of its tools. These programmes will be third party verified in order to ensure internal and external consistency of the information collected.

3.4 Relationship Management and Safeguards

A corporate responsibility collaboration induces a significant change in relationship management among members and towards the suppliers. In fact, the collaboration stimulates the development of an efficient system to share each other's information (concerning the suppliers as well as innovative inputs and mutual suggestions). Among the participants, communication is reasonably open: partners share information required to make the suppliers' relationship work, including their objectives/goals, technical data, information about conflicts, trouble spots or sourcing changing situations. They also share other resources (e.g. materials and technology) and best practices to increase the efficiency and effectiveness of supply chain management at an international level. The sharing of knowledge across organisational boundaries aims to alleviate problems of information asymmetry and ensures convergence in learning and operational skills (Appolloni et al. 2013). A consistent portion of the time spent by the coalition members together is dedicated to regular meetings in which the terms of collaboration are continuously discussed for improvement. Such meetings are also the arena in which the safeguards against opportunistic behaviour are conceived and enforced.

Safeguards—or protective mechanisms—are basically "means to ensure that an agent fulfils the obligations according to the agreement" (Koch 1995: p. 8). The common rationale for safeguards is that "the agent functions as a utility maximiser and the safeguards therefore all contain some aspect of formal, standardised and foreseeable economic incentive, either positive or negative" (Eriksen and Jessen 2009: p. 32). The safeguards can lead to compliance because they influence the incentives of the suppliers so that it is in their best economic interests to comply. Thus, safeguards are a means to reduce opportunism by making it economically wise not to behave opportunistically and by making the principal able to impose serious economic consequences on the agent (Buvik and Reve 2002).[2]

[2]From this, it follows that the contractual control is especially usable in buyer–supplier relationships that are not necessarily long-term, where power rests mainly with one side of the relationship, desired output is clear and measurable and incentives can directly be connected to the output (Tomkins 2001).

Adapting prior studies (e.g. Eriksen and Jessen 2009; Pedersen and Andersen 2006), we consider two different safeguards which can ensure compliance with CSR strategy in the global supply chain: (1) reward and sanction and (2) reputation effect. With regard to each safeguard, the impact of corporate responsibility collaboration will be analysed.

3.4.1 Rewards and Sanctions

Rewards and sanctions help reduce the agency problem of conflicting goals. In the case of CSR compliance, they relate to the cost/benefit imbalance inherited. Two different utilisations of rewards exist. The first is to put a premium on products bought by compliant suppliers. The second is to help the supplier with increased costs related to their ability to comply. However, such joint investments need to be a consequence of compliant behaviour in order to function properly. A sanction is the function opposite to a reward and it is developed to penalise non-compliance. The fastest and most ultimate sanction is to break off the relationship with the supplier in the case of non-compliance. Considering the above, one might say that collaboration among buyers has the potential to enforce both the rewarding and sanctioning safeguards. On the one side, the collaborating buyers—which are sharing objectives and practices of the CSR strategy—have more resources and possibilities to recompense compliant suppliers. Besides, the collaborating buyers might jointly develop more efficient practices to help the supplier with the increased costs related to the CSR implementation. On the other side, their collective efforts might also result in a more effective and efficient sanctioning system for the suppliers. Through an increase in the bargaining power of buyers and in the supplier dependence on all of them, a collective CSR strategy makes reward and sanction safeguards more credible signals to suppliers. Because their behaviour would be analysed by a multiplicity of buyers—all acting together with a relevant economic importance—the suppliers have a strong incentive to honour the shared standards and all the other CSR practices.

3.4.2 Reputation Effect

Reputation effect is seen as a resource that influences future income (Koch 1995). Reputation is also the main reason why companies are adopting codes of conduct in the first place (Pedersen and Andersen 2006). This protective mechanism acts in two ways. On the one hand, the company's efforts to be seen as a socially responsible and reliable transaction partner positively impact on the motivation of the suppliers to become part of the relationship and thus to comply with the shared practices. On the other hand, the reputation of the supplier can be seen as a resource, which influences future economic transactions (Bensaou and Anderson 1999). For instance, if a supplier acts opportunistically, the buyer will probably not engage in future contracts. Moreover, the buyer might indicate to other companies that the supplier is an unreliable partner. The supplier's choice of action therefore depends on a valuation of the costs and benefits of being a reliable partner.

The corporate sustainability collaboration impacts on both the functions of reputation. In fact, the development of a coalition for sustainability by a multiplicity of buyers demonstrates their commitment to become socially responsible partners both towards their customers and suppliers. The establishment of a stable coalition contributes to the buyers' reputation as reliable and honest partners as it proves a true commitment to the sustainable development of economic exchange. As a consequence, membership of a coalition makes the buyers more attractive for the suppliers—due to their true commitment and engagement in sustainability practices—and increases the motivations for the suppliers to comply with the buyers' requirements. The corporate responsibility collaboration might also reduce the costs associated with establishing and maintaining a reputation as an honest and trustworthy partner for all the buyers involved, thus increasing the motivational role and utility of this safeguard for suppliers' compliance. At the same time, as the applicability of the reputation effect depends on the supplier's overall interest in keeping a good relationship with the buyer, a coalition for sustainability might amplify this dynamic by increasing the number of buyers involved, which are all relevant partners for the supplier. In this case, reputation is a relevant safeguard because the supplier behaviour can influence future transactions with a multiplicity of important buyers, and/or a single buyer can harm the supplier by communicating non-compliance to other buyers (coalition members) or to other relevant stakeholders. This could have a stronger resonance in the business network than in the case of a one to one buyer–supplier relationship.

As suggested by prior studies (e.g. Delbufalo and Bastl 2018; Koch 1995; Pedersen and Andersen 2006), the safeguards' enforcement consequent to the development of a corporate responsibility coalition increases the protection of the buyers from the threat of non-compliance by suppliers. In other words, the coalition for sustainability enforces the safeguarding mechanisms and thus reduces the likelihood of non-compliance by the suppliers. Box 5 reports additional details about safeguards in JAC and SAC experiences.

Box 5—Relationship management and safeguards in JAC and SAC

One of the main objectives of collaboration both in the case of JAC and SAC is to obtain reliable and high-quality information at a lower cost. An increase in transparency allows the coalitions' members to develop more efficient relationship management strategy oriented not only to detect suppliers' opportunistic behaviour—and employing corrective mechanism to compensate the suppliers accordingly—but also to help the suppliers to understand the sustainability impact on and benefits to their processes. The collaboration acts in a way that discourages the suppliers from non-compliance: on the one hand, the suppliers find it more difficult to cover up and dissimulate their performance due to the improved monitoring system; on the other hand, the open communication among the coalition members increases the possibility for dissemination of the results concerning the suppliers' behaviour, thus increasing the resonance effect of bad actions.

With regard to sanction and reward mechanisms, the way rewards work under JAC collaboration is linked to the results of the auditing process. JAC members developed several practices which are participated by all buyers. In case of compliance, JAC members neither put a premium on products bought by compliant suppliers nor help the suppliers with the increased costs related to their ability to comply. However, JAC members have developed several practices to transfer innovative input towards the compliant suppliers, thus enhancing their competitive value. Furthermore, each JAC member is encouraged to increase the economic transactions and the sourcing volume with the compliant suppliers. As a result the suppliers, which have been recognised as reliable and socially responsible by one buyer, have the possibility to start new transactions with the other JAC members, finally improving the suppliers' competitive position. The way direct sanctions work under JAC collaboration is that each supplier is audited against the shared standards by the auditing companies. As JAC members have a strategy of engaging in long-term relationships with the suppliers, they do not break off relationships due to non-compliance with the CSR requirements as long as the suppliers show a willingness to improve conditions. Instead, they require that the suppliers prepare a written action plan detailing how the non-compliance issue will be rectified. The buyers require that the suppliers carry out the corrective actions within a short time period (often just four weeks) (Pedersen and Andersen 2006). If a supplier is not able to fulfil the JAC members' requirements, their future relationships with all the buyers would be subject to serious discussion and their current economic transactions terminated. JAC members' ability to use contract termination as a safeguarding mechanism should in large part be seen in the light of their collective economic importance for the suppliers. Due to JAC's large size, its members together constitute important customers for many of the Asian suppliers. As a result of their dependence on JAC buyers as customers, many suppliers are often willing to go to great lengths to fulfil buyers' requirements. They simply cannot afford to give the standards a low priority, as this might imply a termination of the relationship with all the JAC members (see for comparison Pedersen and Andersen 2006).

In the case of the SAC coalition, reputation has proved to be an efficient safeguarding mechanism. In this setting, this protective mechanism operates in a double way. On the one hand, the development of the SAC project aims to improve the environmental and social performance for all members as well as optimise CSR assessments and follow-up processes due to the exchange of best practices; in doing so, this collaboration among competitors also enhances corporate image and reputation for all the members. In the eyes of many suppliers, SAC members are considered "tough" customers because of their commitment to CSR issues and their adoption of accredited tools (Higg Index) to monitor and manage their operations. The enhancement of corporate image and reputation of SAC members impacts on the motivation of suppliers to become part of the relationship with them and thus to comply with the coalition requirements. In fact, SAC members have experienced that several suppliers consider compliance with their sustainability approaches and tools to be a good reference for other (potential) customers, and consequently they are willing to go to great lengths to fulfil the requirements. On the other hand, the reputation effect acts as a safeguard with regard to the diffusion of information about suppliers' behaviour. In fact, the high degree of interconnection among the SAC members discourages suppliers from considering opportunistic behaviours, even for just a single situation. In the case of collaboration among buyers, the costs of being non-compliant would be much more than in the case of single buyer–supplier relationships and straightens the protection provided by the reputation safeguard. Here, reputation is a relevant safeguard because the supplier behaviour can influence future cooperation with a multiplicity of important buyers.

References

Aguinis, H., & Glavas, A. (2012). What we know and don't know about corporate social responsibility: A review and research agenda. *Journal of Management, 38*(4), 932–968.

Amaeshi, K. M., Osuji, O. K., & Nnodim, P. (2008). Corporate social responsibility in supply chains of global brands: A boundaryless responsibility? Clarifications, exceptions and implications. *Journal of Business Ethics, 81*(1), 223–234.

Andersen, M., & Skjøtt-Larsen, T. (2009). Corporate social responsibility in global supply chains. *Supply Chain Management: An International Journal, 14*(2), 75–86.

Appolloni, A., Risso, M., & Zhang, T. (2013). Collaborative approach for sustainable auditing of global supply chains. *Symphonya, 2*, 19–31.

Arrow, K. (1985). The economics of agency. In J. Pratt & R. Zeckhauser (Eds.), *Principals and agents: The structure of business* (pp. 37–51). Boston, MA: Harvard University Press.

Arya, B., & Salk, J. (2006). Cross-sector alliance learning and effectiveness of voluntary codes of corporate social responsibility. *Business Ethics Quarterly, 16*(2), 211–234.

Barrientos, S. (2002). Mapping codes through the value chain. In R. Jenkins, R. Pearson, & G. Seyfang (Eds.), *Corporate Social responsibility and labour rights—Codes of conduct in the global economy* (pp. 61–76). London: Earthscan Publications Ltd.

Bensaou, M., & Anderson, E. (1999). Buyer supplier relations in industrial markets: When do buyers risk making idiosyncratic investments? *Organization Science, 10*(4), 460–481.

Bergen, M., Dutta, S., & Walker, O. C., Jr. (1992). Agency relationships in marketing: A review of the implications and applications of agency and related theories. *The Journal of Marketing, 56*(3), 1–24.

Buvik, A., & Reve, T. (2002). Inter-firm governance and structural power in industrial relationships: The moderating effect of bargaining power on the contractual safeguarding of specific assets. *Scandinavian Journal of Management, 18*(3), 261–284.

Ciliberti, F., de Haan, J., de Groot, G., & Pontrandolfo, P. (2009). Codes to coordinate supply chains: SMEs' experiences with SA8000. *Supply Chain Management: An International Journal, 14*(2), 117–127.

Das, T. K., & Rahman, N. (2002). Opportunism dynamics in strategic alliances. In F. J. Contractor & P. Lorange (Eds.), *Cooperative strategies and alliances* (pp. 89–118). Kidlington Oxford: Elsevier Science Ltd.

Delbufalo, E., & Bastl, M. (2018). Multi-principal collaboration and supplier's compliance with codes-of-conduct. *International Journal of Logistics Management* (forthcoming).

Eisenhrdt, K. M. (1989). Agency theory: An assessment and review. *Academy of Management Review, 14*(1), 57–74.

Emmelhainz, M., & Adams, R. (1999). The apparel industry response to 'sweatshop' concerns—A review and analysis of codes of conduct. *Journal of Supply Chain Management, 35*(3), 51–57.

Eriksen N. S., & Jessen Z. K. (2009). *Enhancing code of conduct compliance. Exploring SME possibilities of control among Chinese suppliers* (Masters Thesis). Department of Management, Copenhagen Business School, 12 May.

Gereffi, G. (2001). Beyond the producer-driven/buyer driven dichotomy. The evolution of global value chains in the internet era. *IDS Bulletin, 32*(3), 30–40.

Grayson, D. (2014, August 4). Why collaborating with the competition can make business sense. *The Guardian*. https://www.theguardian.com/sustainable-business/collaborating-competition-business-sense-sustainability.

Greve, C. (2000). Exploring contracts as reinvented institutions in the Danish Public Sector. *Public Administration, 78*(1), 153–164.

Hill, R. P., & Rapp, J. M. (2014). Codes of ethical conduct: A bottom-up approach. *Journal of Business Ethics, 123*, 621–630.

Jenkins, R. (2001). *Corporate codes of conduct. Self-regulation in a global economy*. Geneva: United Nations Research Institute for Social Development.

Jensen, M. C., & Meckling, W. H. (1976). Theory of the firm: Managerial behavior, agency costs, and ownership structure. *Journal of Financial Economics, 3*(4), 305–360.

Jiang, B. (2009). The effects of interorganizational governance on supplier's compliance with SCC: An empirical examination of compliant and non-compliant suppliers. *Journal of Operations Management, 27*, 267–280.

Joint Audit Cooperation (JAC). (2017). *Audit process overview. Information for suppliers. Version 01-20140116*. Available at http://jac-initiative.

Kapstein, E. B. (2001). The corporate ethics crusade. *Foreign Affairs, 80*(5), 105–119.

Koch, C. A. (1995). *Economic transactions, opportunistic behavior and protective mechanisms*. Odense, Denmark: University of Southern Denmark, Department of Management.

Kolk, A. (2000). *Economics of environmental management*. Edinburgh Gate, Harlow, England: Pearson Education Limited.

Lassar, W. M., & Kerr, J. L. (1996). Strategy and control in supplier-distributor relationships: An agency perspective. *Strategic Management Journal, 17*(8), 613–632.

Lee, K.-H., & Kim, J. W. (2009). Current status of CSR in the realm of supply management: The case of the Korean electronics industry. *Supply Chain Management: An International Journal, 14*(2), 138–148.

Maloni, M. J., & Brown, M. E. (2006). Corporate social responsibility in the supply chain: An application in the food industry. *Journal of Business Ethics, 68*(1), 35–52.

Mele, V., & Schepers, D. H. (2013). E pluribus unum? Legitimacy issues and multi-stakeholder codes of conduct. *Journal of Business Ethics, 118*, 561–576.

Milgrom, P., & Roberts, J. (1992). *Economics, organization and management*. New Jersey: Prentice Hall Inc.

Mishra, D. P., Heide, J. B., & Cort, S. G. (1998). Information asymmetry and levels of agency relationships. *Journal of Marketing Research, 35*(3), 277–295.

Mitnick, B. M. (1973). *Fiduciary rationality and public policy: The theory of agency and some consequences. 1973 Annual Meeting of the American Political Science Association*. New Orleans, LA: Proceedings of the American Political Science Association.

Network for Business Sustainability (NBS). (2017). *Competitor collaboration is more than a trend* (September 8). https://nbs.net/p/competitor-collaboration-is-more-than-a-trend-21579740-5a40-43d2-80f1-c3475ace1738.

O'Donnell, S. W. (2000). Managing foreign subsidiaries: Agents of headquarters, or an independent network? *Strategic Management Journal, 21*(5), 525–548.

Pedersen, E. R., & Andersen, M. (2006). Safeguarding corporate social responsibility (CSR) in global supply chains: How codes of conduct are managed in buyer-supplier relationships. *Journal of Public Affairs, 6*(3/4), 228–240.

Peloza, J., & Falkenberg, L. (2009). The role of collaboration in achieving corporate social responsibility objectives. *California Management Review, 51*(3), 95–113.

Perrow, C. (1986). *Complex organizations: A critical essay* (3rd ed.). New York: Random House.

Petersen, T. (1993). The economics of organization: The principal-agent relationship. *Acta Sociologica, 36*(3), 277–293.

Prosman, E. J., Scholten, K., & Power, D. (2016). Dealing with defaulting suppliers using behavioral based governance methods: an agency theory perspective. *Supply Chain Management: An International Journal, 21*(4), 499–511.

Risso, M. (2012). A horizontal approach to implementing corporate social responsibility in international supply chains. *International Journal of Technology Management, 58*(1/3), 64–82.

Saam, N. J. (2007). Asymmetry in information versus asymmetry in power: Implicit assumptions of agency theory? *The Journal of Socio-Economics, 36*, 825–840.

Sethi, S. P. (2002). Standards for corporate conduct in the international arena: Challenges and opportunities for multinational corporations. *Business and Society Review, 107*(1), 20–40.

Sobczak, A. (2006). Are codes of conduct in global supply chains really voluntary? From soft law regulations of labour relations to consumer law. *Business Ethics Quarterly, 16*(2), 167–184.

Tomkins, C. (2001). Interdependencies, trust and information in relationships, alliances, and networks. *Accounting, Organizations and Society, 26*, 161–191.

Utting, P. (2000). *Business responsibility for sustainable development*. United Nations Research Institute for Sustainable Development (UNRISD), Geneva.

Whipple, J. M., & Roh, J. (2010). Agency theory and quality fade in buyer-supplier relationships. *The International Journal of Logistics Management, 21*(3), 338–352.

Williamson, O. E. (1993). Calculativeness, trust and economic organization. *Journal of Law and Economics, 36*, 453–486.

World Bank. (2003). *Strengthening implementation of corporate social responsibility in global supply chains*. The World Bank.

Bibliography

Bitch Olsen, P., & Pedersen, K. (1999). *Problemorienteret pro-jektarbejde: En værktøjsbog* (2nd ed.). Copenhagen: Roskilde Universitetsfor-lag.

Castka, P., & Balzarova, M. A. (2008). ISO 26000 and supply chains—On the diffusion of the social responsibility standard. *International Journal of Production Economics, 11*, 274–286.

Dees, J. G. (1992). Principals, agents and ethics. In N. E. Bowie & R. E. Freeman (Eds.), *Ethics and agency theory* (pp. 25–59). New York/ Oxford: Oxford Uni-versity Press.

Eltantawy, R. A., Giunipero, L., & Fox, G. L. (2009). A strategic skill based model of supplier integration and its effect on supply management performance. *Industrial Marketing Management, 38*(8), 925–936.

Fama, E. F., & Jensen, M. C. (1986). Separation of ownership and control. *Journal of Law and Economics, 26,* 327–349.

Gereffi, G. (1994). The organisation of buyer-driven global com-modity chains: how US retailers shape overseas production net-works. In G. Gereffi & M. Korzeniewicz (Eds.), *Commodity chains and global capitalism* (pp. 95–122). CT: Greenwood Press, Westport.

Jenkins, R., Pearson, R., & Seyfang, G. (2002). Introduction. In R. Jenkins, R. Pearson, & G. Seyfang (Eds.), *Corporate respon-sibility and labour rights. Codes of conduct in the global economy* (pp. 1–10). London: Earthscan Publications.

Kortelainen, K. (2008). Global supply chains and social re-quirements: Case studies of labour condition auditing in the Peo-ple's Republic of China. *Business Strategy and the Environment, 17*(7), 431–443.

Miles, M., & Huberman, A. M. (1994). *Qualitative data analysis*. Thousand Oaks, CA: Sage Publications.

Müller, M., Gomes dos Santos, V., & Seuring, S. (2009). The contribution of environmental and social standards towards ensuring legitimacy in supply chain governance. *Journal of Business Ethics, 89*(4), 509–523.

Neergaard, P., & Pedersen, E. R. (2003). Corporate social behav-iour. Between the rules of the game and the law of the jungle. *Journal of Corporate Citizenship, 12,* 43–57.

Patton, M. Q. (2002). *Qualitative research & evaluation methods* (3rd ed.). London, UK: Sage Publications.

Saunders, M., Lewis, P., & Thornhill, A. (2003). *Research meth-ods for business students* (3rd ed.). Essex, England: FT Prentice Hall.

Simatupang, T. M., & Sridharan, R. (2005). The collaboration in-dex: A measure for supply chain collaboration. *International Journal of Physical Distribution & Logistics Management, 35*(1), 44–62.

Tirole, J. (2006). *The theory of corporate finance*. Princeton University Press.

Voss, C., Tsikriktsis, N., & Frohlich, M. (2002). Case research in operations management. *International Journal of Operations & Production Management, 22*(2), 195–219.

Welford, R., & Frost, S. (2006). Corporate social responsibility in Asian supply chains. *Corporate Social Responsibility and Envi-ronmental Management, 13,* 166–176.

Wong, C. Y., & Boon-itt, S. (2008). The influence of institutional norms and environmental uncertainty on supply chain integration in the Thai automotive industry. *International Journal of Production Economics, 115*(2), 400–410.

Yin, R. K., (2008). *Case study research: Design and methods* (4th ed.). Sage Publications Inc.

Kolibri 360 GmbH
Ferdinand-Jühlke-Straße 7
99095 Erfurt, Deutschland
produktsicherheit@kolibri360.de